Moral Neuroeducation for a Democratic and Pluralistic Society

Patrici Calvo • Javier Gracia-Calandín
Editors

Moral Neuroeducation for a Democratic and Pluralistic Society

Springer

Editors
Patrici Calvo
Universitat Jaume I
Castellón de la Plana, Spain

Javier Gracia-Calandín
Universidad de Valencia
Valencia, Spain

ISBN 978-3-030-22561-2 ISBN 978-3-030-22562-9 (eBook)
https://doi.org/10.1007/978-3-030-22562-9

This Springer imprint is published by the registered company Springer Nature Switzerland AG.
The registered company address is: Gewerbestrasse 11, 6330 Cham, Switzerland

Acknowledgements

The contents of this book form part of the wide-ranging study of ethics and neuroscience that has been undertaken over the course of the last 9 years in the form of three nationally funded research projects, which have been financed by the Ministerio de Ciencia e Innovación (Ministry for Science and Innovation), the Ministerio de Economía y Competitividad (Ministry for Economy and Competitiveness), and Ministerio de Ciencia, Innovación y Universidades (Ministry for Science, Innovation and Universities): "La ética del discurso ante los retos de la neuroética y la neuroeconomía" (Discourse ethics and the challenges posed by neuroethics and the neuroeconomy) (FFI2010-21639-C02), "Racionalidad práctica en perspectiva neuroética" (Practical rationality from the perspective of neuroethics) (FFI2013-47136-C2) and "Neuroeducación moral para una sociedad pluralista y democrática" (Moral neuroeducation for a pluralistic and democratic society) (FFI2016-76753-C2).

Over the duration of these three projects, we have received a wealth of support and advice from numerous individuals, as well as assistance and funding from a range of public and private institutions. First and foremost, we would like to express our sincere appreciation to Jesús Conill, Adela Cortina and Domingo García-Marzá for managing, guiding and contributing to these projects over the course of the last 9 years. Second, we would like to express our gratitude to the Universidad de Valencia and the Universitat Jaume I de Castelló for their institutional support, as well the Spanish Government for the generous financial support that has enabled us to undertake these three research projects from 2010 to 2019. Finally, we would like to thank the Turisme Comunitat Valenciana, the Instituto Valenciano de Tecnologías Turísticas (Invat.tur), the Autoridad Portuaria de Castellón and Unión de Mutuas for their contribution and assistance, as well as UBE Corporation Europe (UCE) and Balearia for their role in testing, applying and implementing the knowledge gathered over the course of these projects.

Introduction

The Moral Neuroeducation: A Field to Discover

Over the last 20 years, the addition of the prefix *neuro* to a range of disciplines and subdisciplines within the humanities and social sciences has become one of the most widely debated recent scholarly developments. Two principal concerns underpin this recent phenomenon, although they are by no means the only ones. On the one hand, there is the powerful appeal of neuroscience, the findings of which have managed to seduce a range of representatives from the professional and technological world—politicians, businesspeople, advertising agents, academics, etc.—as well as the media and general public. On the other hand, there is the belief that all knowledge based on a neuroscientific foundation, thanks to the application of neuroscientific research methods and techniques, and the analysis of images of cerebral activity, acquires the status of a natural science capable of both describing reality and predicting behaviour patterns and their consequences and which is, furthermore, able to control and dominate human behaviour through the observation and mathematisation of and experimentation on observable, measurable and verifiable behavioural facts. We are thereby confronted by a trend towards the *neurodatafication* of human behaviour, or, in other words, the gathering and processing of cerebral data concerning peoples' behavioural activity related to a range of activities in order to develop a greater knowledge and deeper understanding of the laws that govern this behaviour while also improving the processes involved and the benefits that can be attained.

The sphere of education has not been an exception from this trend. Although there are precedents in this sense, such as the proposal made by the neurodidact Gerhard Preiss in 1988 and the use of the concept by Mary Catherine Eakin Dozier in her study "A Case Study of a Curriculum Development Effort", the widespread inclusion and use of the prefix *neuro* in the realm of education in a contemporary neuroscientific sense did not take place until 2002, in the wake of the OECD report *Understanding the Brain: Towards a New Learning Science* (2002), which was published long after the emergence of the subdisciplines such as neuropolitics, neuroeconomy, neuropsychology and neuroethics, in addition to many others. However, in the last few years, this new field of knowledge had grown exponentially, both with regard to academic output and the interest shown by those sectors that sought to apply it, such as business, management, politics, health, artificial intelligence and of course the diverse realms of education.

Neuroeducation is a new neuroscientific discipline that has emerged from the synergic convergence of three areas of knowledge: neuroscience, education and psychology. Its principal aim is to take advantage of their combined findings in order to increase our knowledge of the cognitive foundation of the human brain, to identify the pedagogic premises involved and, finally, to improve both teaching methods and learning processes and the curricula involved in their correct development.

In this sense, during the last decade, neuroscience has made a significant contribution to the increase of theoretical and practical knowledge regarding all these issues, above all in relation to the identification of the cerebral foundation of human cognition, the comprehension of the elements and processes involved in learning and the development and improvement of techniques and strategies for their normative implementation, as well as practical application. Concerning this range of developments, various issues relevant to teaching-learning processes are worthy of note: the role of mirror neurons, the importance of brain plasticity, the significance of the emotions and the possibility of acquiring new knowledge and information through repetition exercises.

Firstly, mirror neurons are responsible for a major part of human behaviour, from the very first months of life onwards. For the purpose of adaptation, they form emotional connections between the brains of those who belong to a specific society in order to generate behavioural mimesis that is based on evaluations of direct experience. Within this vicarious learning process, feelings and emotions play a fundamental role as they allow members of a society to better adapt themselves through unconscious learning processes, which are much quicker than the conscious ones that use imitation and empathy.

Secondly, the human brain is highly malleable. In other words, by strengthening or weakening the synapses connecting neurons, and through neurogenesis, the nervous system has the capacity to change the brain's structure and function in order to continually adapt to its environment, experience and stimuli. For education, this neural plasticity implies the possibility of being able to influence the cognitive development of any student to improve their capacities and eliminate learning disorders that hinder the development of their abilities or make learning impossible.

Thirdly, emotions have a cognitive basis and are extremely important in all learning, reasoning and decision-making processes. As important neuro-educators have argued, positive emotions encourage memorisation and learning, while negative emotions make the transfer of information between the different areas of the brain and the effective processing of that information more difficult.

Finally, the brain is capable of forging links between the knowledge acquired and new information by repeating what it wishes to assimilate, such as values, information, guidelines, behaviour, etc., whereby, the possibility arises for learning to be optimised and for critical reflection to be fostered among students. The brain's capacity for link creation is, among other things, due to the automation of mental processes, which frees up the working memory for other uses, and, as has been shown, the more working memory that is available to an individual, the better prepared they are for learning and critical reflection.

Despite acknowledgement of the importance of these and other advances in our understanding of teaching-learning processes, these contributions, their possible applications and their theoretical and practical outcomes have also been subject to criticism. For example, the recent special issue "Neuroética y Neuroeducación: repensando la relación entre las neurociencias y las ciencias sociales" [Neuroethics and Neuroeducation: rethinking the relationship between neuroscience and the social sciences], edited by Daniel Pallarés-Domínguez and Andrés Richart for the journal *Recerca. Revista de Pensament i Anàlisi*, examined the major controversy about whether neuroscientific advances in the field of education enable the emergence and development of societies that are more just and enjoy a higher degree of happiness. Or, in other words, does educating individuals in such a way that they develop social competences, such as empathy and prosocial emotive traits, as well as intellectual faculties and executive functions, such as the use of reason and memory, suffice to achieve the level of postconventional development demanded by pluralist and democratic societies. Among other things, as has been demonstrated, over the course of the brain's evolutionary process, the human brain has incorporated predispositions that incline it to care and cooperation but also to tendencies that give rise to egotistical, nepotistic, xenophobic, misogynist and aporophobic behaviour. The brain's plasticity enables the cultivation of both classes of predisposition. It is precisely for this reason that it is necessary to acknowledge and develop an in-depth understanding of these tendencies and their consequences for society, as well as to explore to what extent we may wish to and be able to deploy them in a democratic society through moral neuroeducation.

Indeed, this was the dual aim of the combined research and technological development project "Neuroeducación moral para una sociedad pluralista y democrática [Neuroeducation for a pluralist and democratic society]", directed by Adela Cortina, Jesús Conill and Domingo García-Marzá. It was funded by the Ministerio de Ciencia y Competitividad (Ministry for Economy and Competitiveness) between 2016 and 2019, which involved a team of researchers from the Universidad de Valencia, the Universitat Jaume I in Castellón, the Universidad Politécnica de Valencia and the Universidad de Murcia. Firstly, it aimed to propose an ethical—dialogic and cordial—concept for moral neuroeducation, one based on an ethics of a renewed discourse, such as that of the knowledge, provided by neuroscience through practical neurophilosophy, of how the brain functions, which can serve as a model for the design, development and critique of a range of proposals, methods and educational plans. Furthermore, it can help to eradicate or minimise prejudices that persist in society and that hinder and even impede society's correct development, for example, hate speech, corruption, intolerance, nepotism, aporophobia (the rejection of the poor), xenophobia, misogyny and homophobia at both a micro- and macro-level. Secondly, this project sought to offer guidelines and frameworks that can be applied and implemented in both formal and informal education, deliberative politics, business and organisational management, the diverse processes of digital transformation and public policies intended to contribute to human development.

The aim of this book is to address the ideas developed over the last 3 years of this project through contributions made by a selection of its researchers. Of special interest is how and in what sense a dialogic and cordial ethics can prove to be highly fruitful for reinforcing the values that are appropriate to a democratic and pluralist society; for establishing models of human development that empower the most vulnerable; to help overcome prejudices, phobias and corrosive sentiments; and, finally, as a means of outlining models of institutional design that are capable of guiding a range of professional practices in a just and felicitous direction for all parties involved.

The chapters that follow are divided into three parts, the first of which tackles a series of questions regarding the ethical model that should serve as a foundation for moral neuroeducation. This opening part consists of four chapters that analyse the type of ethics deemed necessary for moral neuroeducation. The second part addresses the important role played by the emotions for moral neuroeducation, and the third and final part considers various aspects of the application of moral neuroeducation.

In the opening chapter, Jesús Conill tackles the question of whether the ethics that moral neuroeducation seeks to develop has to be naturalistic. He thereby undertakes an analysis of the concept of naturalisation and warns of the danger of ethics being understood in the reductionist terms of naturalism. While he acknowledges the importance and need for moral philosophy to incorporate a neuroscientific perspective, he also highlights the need to overcome the naturalisation of moral life for one to become fully aware of moral issues. The distinction he draws between education and learning, in conjunction with his proposal of a humanistic neuroeducation, underscores the capital importance of the moral dimension of neuroeducation.

In the second chapter, Javier Gracia-Calandín undertakes a further reflection on the development of a non-naturalistic model of neuroethics for moral neuroeducation. For this purpose, he set outs the overarching goal of education and the limitations of current neuroeducation, such as it being understood in terms of learning modes based on instrumental skills, such as mathematical calculation and reading. Having signalled the limits of a naturalist neuroethics, Gracia-Calandín examines four significant contributions made by moral neuroeducation: combating demoralisation in education, the eradication of social phobias, the cultivation of the values and virtues of civic ethics through education and, finally, the rejection of all forms of moral hypocrisy.

In the third chapter, Andrés Richart presents a series of fundamental contributions for moral neuroeducation from a threefold—phylogenetic, ontogenetic, functional—perspective. Both in the case of phylogenesis and ontogenesis, a number of types of prosocial behaviour become moral under specific conditions. The functional perspective is also necessary to explain the function of morality at both a cerebral and psychological level. Knowledge of the psychological faculties and the cerebral structures of morality, as well as how these function, is clearly relevant for the creation of tools and educational models that teach moral values, norms and principles.

In the fourth and final chapter of the first part, César Ortega-Esquembre analyses the specific ethical models that underpin the task of moral education. Founded on the conceptual matrix of Adela Cortina's ethics of cordial reason and Habermas's discourse ethics, a pressing issue he highlights is the role of "liberal elements". The proposal that he develops is one of the ethics founded on universal principles such as freedom, equality and a respect for pluralism. This ethical model constitutes the best way of avoiding any possible suspicion of indoctrination, which would risk undermining moral neuroeducation.

The second part of the book develops a focus on the key role played by the emotions in moral neuroeducation. It consists of four chapters that explore questions related to feelings, humour, ethical imagination and corporeality.

In Chap. 5, Lidia de Tienda considers the central role played by the emotions in moral education. In regard to this, she challenges a number of assumptions underpinning current neuroscientific methodologies. Firstly, she distinguishes moral education from emotional management; secondly, she argues that the emotions contain four essential components, which render emotions a privileged means of accessing the sphere of values; and, thirdly, she states that moral education ought to focus on methods of practical deliberation that take the emotions into account.

In Chap. 6, Juan Carlos Siurana addresses the importance of neuroeducation for an understanding of ethical humour. In accordance with neuroscientific studies, ethical humour is acknowledged as being beneficial for health, and the degree to which human beings laugh in an ethical or correct manner reflects to what extent they have received an appropriate ethical education. From the perspective of healthcare ethics, Siurana explores key aspects of ethical humour in terms of promoting virtues such as patience, tolerance, kindness, humility, perseverance and valour. He shows how consideration of the contribution made by ethical humour can enable moral neuroeducation to play a fundamental role in benefitting individuals and societies alike.

In the seventh chapter, Francisco Arenas-Dolz discusses how the moral imagination may be gauged through the theory of triune ethics, and he argues that the literary mind is an indispensable tool for everyday reasoning and the moral imagination. Arenas-Dolz also considers the neurological foundations of the imagination and seeks to move beyond reductionist theories in order to include other perspectives.

In the eighth and last chapter of the second part, Marina García-Granero undertakes a critique of the neuroessentialist paradigm that locates subjectivity within the confines of the brain. Taking a Nietzschean focus, she proposes to replace it by developing an overarching concept of the self as based, firstly, in the body, whereby an individual's character is inextricably linked to their emotions, feelings and values, and, secondly, within human reality, which not only pays attention to the brain but also the body as a whole and the lifeworld. Finally, she considers the relevance of the role played by experience of the self as socially situated within the lifeworld and with regard to the conformation of personal subjectivity.

In the third and final part, which consists of five chapters, the application of moral neuroeducation is addressed in relation to business organisations, neuroleadership, the question of gender, the task of overcoming the rejection of poverty and the impossibility of machines' successfully undertaking any form of moral neuroaprenticeship.

In Chap. 9, Elsa González-Esteban examines the contribution of neuroethics, neuroeducation and organisational neuroscience in order to open up spaces for participation and deliberation within organisations, whereby the moral character of the latter may be cultivated in a more effective manner. The spaces she discusses are codes of ethics, social responsibility programmes, sustainability reports and ethics committees.

In Chap. 10, Daniel Pallarés-Domínguez focuses on proactive epigenesis, which is a concern currently being addressed in neuroethics, and he considers how it offers a possible foundation for moral neuroeducation. Proactive epigenesis provides a theoretical framework for the study of the effects that social phenomena such as aporophobia have on the brain. Pallarés-Domínguez underscores the importance not only of the material conditions of poverty but also poverty in moral terms.

In Chap. 11, Sonia Reverter-Bañón turns to the case of gender in neuroeducation. She analyses the concept of sex and gender in order to consider how to overcome the binary prejudices that persist in many neuroscientific studies on sex differences. Given that an acritical concept of gender within neuroeducation could foster political decisions that perpetuate inequalities, Reverter-Bañon proposes a critical framework that allows us to develop research on sex/gender differences within the brain.

In Chap. 12, Maria Medina-Vicent analyses the issue of neuroleadership in relation to the neuroeducational value of diversity. The combination of different leadership styles has become one of the central premises of neuroleadership. Medina-Vicent focuses on the value of diversity; from a broad ethical perspective, one that goes beyond the company per se, she argues for not restricting the benefits of diversity in order to increase of productivity. She then goes on to ask whether the quest for diversity is in fact a continuation of the neoliberal discourse on individualism, self-management and self-discipline.

In the final chapter, Patrici Calvo explores the contemporary trend towards the digitisation of all spheres of human activity—political, economic, healthcare, educational, communications, etc.—in order to undertake a critical study of the possibility of whether machines provided with artificial neuronal networks and artificial intelligence are capable of learning to formulate a criticism of what they encounter and take morally valid decisions.

The contents of this book are not an exhaustive enquiry into the range of issues related to the field of moral neuroeducation; instead, it seeks to contribute to the development of a dimension of neuroeducation that has to date not been studied in depth, and furthermore, it seeks to demonstrate the link between this discipline and others such as ethics, neuroethics, neuropolitics, neuromanagement, neuroleadership and artificial intelligences.

Univesitat Jaume I
Castellón de la Plana, Spain
Patrici Calvo

University of Valencia
Valencia, Spain
Javier Gracia-Calandín

Contents

Part I
Ethics for Neuroeducation

Chapter 1
Must Ethics for Moral Neuroeducation Be Naturalistic?

Jesús Conill

1.1 Understanding Naturalism

The naturalization of ethics is being promulgated in a range of domains, one of which neuroscience is, hence the question stated in this chapter's title. In order to answer this question, we must clarify what is meant by "naturalistic"; is it used in a reductionist sense, or solely to refer to a concern for incorporating a neuroscientific perspective into the study of ethics, and thereby without any intention that it should predominate within philosophical discussion.

Naturalism is increasingly prevalent in contemporary intellectual discourse, above all in the Anglo-American world, where there is a pervasive naturalistic programme. Yet it is also encountered in Europe, where a thinker of the renown of Jürgen Habermas has been won over by today's vogue for naturalism, albeit solely in the form of a "soft naturalism" (Habermas 2003a, b, 2006, 2009; Cortina 2011). Based on contemporary scientific knowledge, this resurgence of naturalism seeks to impose itself on all areas of discourse, thereby becoming a new paradigm. Naturalistic objectification holds sway everywhere from epistemology to ethics. Is this merely a fashion? a quasi-ideology, based on a "scientificist faith"? Or flawed philosophy rather than science? Contemporary naturalism attempts to naturalize traditional philosophical concepts, even moral normativity. It makes recourse to the natural sciences, but more recently, and in the wake of the expansion of physics, biology has gained a special status, firstly with regard to genetics, and now neuroscience. As a result of this trend, contemporary philosophy is becoming naturalized across its theoretical and practical spheres.

J. Conill (✉)
Universidad de Valencia, Valencia, Spain
e-mail: Jesus.Conill@uv.es

P. Calvo, J. Gracia-Calandín (eds.), *Moral Neuroeducation for a Democratic and Pluralistic Society*, https://doi.org/10.1007/978-3-030-22562-9_1

Nonetheless, the diverse and highly varied naturalisms that have emerged do not sufficiently elucidate what they mean by "nature" (Flanagan 2009; García-Ruiz 2018); this is a fundamental issue which is almost always overlooked. All too often every kind of philosophical thought that addresses natural evolution tends to be considered naturalistic, as if the consideration of any concept from the perspective of evolution necessarily leads to its naturalisation, which is clearly an unwarranted exaggeration.

Today, within the field of neuroethics, naturalizing means addressing the contributions of the natural sciences within the framework of a Darwinist approach in order to determine what moral life is. Naturalizing thus means justifying ethical principles in terms of evolutionary advantages or success, and an explanation of their genesis in terms of the human being's biological-evolutionary structure, in this case a neuroscientific structure. Naturalization understood in this sense implies both a denial of universally valid moral principles and the view that a phenomenological perspective is deceptive, an epistemological illusion. Thereby, the naturalistic biological genealogy results in the dissolution of the phenomenological dimension of ethics, and instead morality is understood as a product of evolutionary performance. Morality would at the very most prove to be a useful fiction of the human organism.

With regard to what is termed "weak naturalism", it unintentionally fosters a mistaken interpretation of the role of philosophy, as it conveys the idea that nothing can be known unless provided by the empirical approach of the natural sciences. This reductionist doctrine is the most salient feature of contemporary naturalism (Goldman 1995, 518; Müller 2004, 89); it denies philosophy any objectivity and advocates a dedication to the natural sciences (such as the current interest in neurology), instead of philosophy. Naturalism thus favours a scientisitic image of both the world and life, one that clashes with the traditional concepts of life and how they are understood. As a result, it leads to a reconsideration of the relationship between science and philosophy, and above all the consideration of whether the moral life can be fully explained by the methodology of the natural sciences.

However, the moral life can only be addressed in-depth with regard to "the mode of accomplishment" (*im Vollzugsmodus*) (Habermas 2009, 233). To formulate knowledge on the sense of obligation, an empirical observer's perspective does not suffice, as it only informs us about whoever observes. Objectification constitutes a reified abstraction, by means of which the mind is naturalized, its mental operations reduced to cerebral ones. The world of the mind is bound to its own organic substrate, but this bond does not imply it being reduced to this. There are limits to naturalistic self-objectification (Habermas 2009, 262–263).

The naturalizing approach must be overcome through hermeneutics,[1] because the mind's historical-natural (*naturgeschichtlich*) genesis is not merely the result of natural evolution. It is instead a process of cognitive, technical and moral learning, one that is irreducible to natural biological-evolutionary processes, as the former

[1] See Conill (2015) and (2019) for a bio-hermeneutic concept of human nature.

have a transbiological and trans-evolutionary potential, as is illustrated, for example, by the post-conventional development of the moral conscience[2] (Cortina 2016; Ortega-Esquembre 2016), which is so fundamental for education.

1.2 Overcoming the Naturalization of Moral Life

Ortega y Gasset criticised the scientific utopianism of naturalism by referring to the following statement by the physiologist, Loeb: "The day will come when what we today call mankind's moral acts are simply explained as tropisms" (Ortega y Gasset 2006, VI, 55). Given that the "human aspect" of human life eludes the perspective of naturalism: "all the seemingly inexhaustible portents of the natural sciences will always be halted when faced by the strange reality that is human life", because "a human being is not a thing". The natural sciences are unequipped to discuss the human realm in this way. The thought devoted to human life must be undertaken according to categories that are radically different to those used to explain material phenomena. However, this is a challenge, because, according to Ortega y Gasset, for the last three centuries physicism has grown accustomed to setting aside the strange reality that is human life.

In order to comprehend what it means to be human a story has to be told, and this even applies to the endeavour to comprehend the particular dynamics of an individual brain; life only gains a degree of transparency through *historical reason*[3] (Ortega y Gasset 2006, VI, 71). "The "natural" being is always there, underneath the changing historical individual (…,) enduring" (Ortega y Gasset 2006, VI, 324). Yet Ortega y Gasset wonders whether the purpose of vitality is not "the increase of itself, its empowerment and growth, that which mysteriously appears in some of the Greeks' texts under the name of *pleonexía,* fullness, being more". Furthermore, "to live is not to survive, strive to be (…), instead to live is to live more, tending towards plenitude, the will to power—according to Nietzsche" (Ortega y Gasset 2004, II, 583–584, 2006, VI, 73).[4] Ortega y Gasset's explicit reference to Nietzsche signals how he will go on to develop his understanding of the meaning of life and education.

With regard to human life, biological purpose is one matter and the meaning of life another, natural biology is one thing and a life lived something quite distinct; the meaning of life is an ethical issue, "mankind's greatest problem". It is an issue that is resolved in a diverse manner through the utilitarian concept, which Ortega y Gasset and Machado termed as "perverse",[5] and that pulses within Darwinism, and secondly, Ortega y Gasset'S ethical concept of ratio-vitalism. Ortega y Gasset, like

[2] See also Kohlberg (1981), Apel (1973) and Gozálvez (2000).

[3] The brain is also historical and narrative. See Marina (2011).

[4] It is curious to note that at the start of the OECD Report on neuroeducation reference is also made to Nietzsche.

[5] See Machado (2004).

Nietzsche, is opposed to Darwinist naturalism, because this limits the meaning of life on the basis of a reductionist sense of biological purpose. For Darwinism, "the vital purpose", "the purpose of life", "the essential concern of life", is "adaptation to the physical environment". "Darwinism assumes that adaptation to the environment is the purpose that governs all vital phenomena. Living means adapting". "In contrast to that point of view", for Ortega y Gasset, life is "an artificial creative attitude that consists of the increase of one's own being, of its attaining plenitude" (Ortega y Gasset 2004, II, 589–590). To achieve this will require contributions from both endeavours to reform mankind, and the creative imagination that makes these possible (Conill 1991). What is at stake is what life consists of: adaptation or plenitude. In line with the Nietzschean approach and with regard to its decisive sense for education, Ortega y Gasset provides an alternative, which seeks to overcome Darwinist naturalism by offering an enriching understanding of life, in which creation, and an aspiration to excellence extend the horizon of existence.

"The human being is an infinitely malleable entity of which one may make what one will", because "of itself it is nothing, save the mere potential to be"[6] (Ortega y Gasset 2006, VI, 66), albeit within a set of limited possibilities. Nonetheless, Ortega y Gasset asks what those "limits" are, and where the "frontiers" of "human malleability" are, because they are not easy to distinguish, given that individuals discover themselves to be the "mere potential to be", or to become anything by way of education.

Undoubtedly, what most distinguishes the *human condition* is a person's way of "being in reality", which founds being upon a world of possibilities (Ortega y Gasset 2010, X, 3–14, 20–35). Mankind is understood as beings who have fled nature, an "ill-adapted animal", who nature has not equipped with the means to satisfy their own desires and for this reason they revolt against nature; this is what constitutes mankind's "dignity". The concept of mankind that departs from these ideas moves beyond the limits of naturalism, because the fundamental meaning of Ortega y Gasset's concept is dramatic, both narrative and historical; Ortega y Gasset like Goethe (Ortega y Gasset 1983, 231–237; 2009, IX) understood that human life consists of becoming what one must become through education. Naturalism tends to have a "closed" conception of nature (Ortega y Gasset 2010, X, 175–181), but Ortega y Gasset thinks in terms of the possibility of a "historical consistency", which in my view, forges a crucial distinction between a closed and an open structure of reality, between a closed essence and an open one, such as that of the reality of human life, which consists of an "event of a dramatic nature" (Ortega y Gasset 2006, V, 374). It is for this the reason why naturalism is unable to understand the "more strictly human" (Ortega y Gasset 2010, X, 175–181) problems, such as that of the human individual's education.

[6] Today's neuroscience also stresses the brain's "plasticity" as a basis for its creative capacity.

1.3 What Is Meant by "Education"?

Having clarified what is meant by naturalization, the question of education needs to be tackled, because many studies carried out in the new field of neuroeducation tend to equate education with the learning process. In fact, the great majority of publications on neuroeducation deal with the neural bases of understanding language and arithmetic, dyscalculia, dyslexia and learning and language development disorders (Sousa 2011; Battro et al. 2008).

Education and learning are not, however, the same thing. Since ancient times men and women have considered themselves to be educatable beings, both in general and in the specifically moral sense. The Greek *paideia* presupposed the natural capacity to shape oneself, to form an *êthos*, a character fashioned according to the virtues or excellences (*aretai*) deemed most fitting for mankind, and the way of life that leads to happiness (*eudaimonía*). According to this educational approach, based on a specific concept of human nature, the human being is an animal whose own nature (*phýsei*) comprises *lógos* (reason, language and speech). The practical *lógos* (concerning action) encompasses an individual's capacity to choose (*proaíresis*) from the range of possibilities – of what is feasible – the means to an end, or the most valued good, that which is no longer a means to any other end: the happy or fulfilling life.

If we really knew what such a life consisted of, we would have to educate people to be happy, to become happy. Such a eudemonist concept of education is nevertheless confronted by at least two major issues. The first of these concerns the possibility of a knowledge of human nature, as an education in happiness would have to be based on this knowledge. The second issue stems from the communal nature of virtue, or in other words, the communal contextualism upon which the traditional theory of virtue is based on, and that steers education towards the "common good" of a specific civic community. Regrettably, such a particularist, even "political" (*pólis*), contextualism has the disadvantage of not standing firm against the critique of the modern Enlightenment. The neurosciences could nevertheless help to develop a knowledge of human nature, and reveal some of the universalizable virtues (valid for different cultures), which would be useful as a guide for neuroeducation (Marina 2011).

The modern Enlightenment also upheld the possibility of perfecting mankind in the light of reason, despite the difficulties entailed by such an endeavour, as stated by Kant himself in *On Education* (1983),[7] in which he stressed that the most difficult problem for genuine human progress is that of education. Yet, in contemporary life the challenge is to identify a project for "mankind's reform" that guarantees moral progress. And this is accentuated in the new socio-historical context in which the growing power of techno-science is leading to the emergence of a utopian horizon, one defined by the new technologies, and that is filling human life with new symbols, expectations and hopes. In these times, characterised by the boom of the technocratic

[7] See Cortina (2011).

realm, we continue to face the same problem confronted, in a religious sense by Saint Paul, in a scientific-social sense by Marx, and by Nietzsche in a poetic-philosophical sense; a problem that nobody has solved: is "human nature" morally improvable and perfectible? However, this utopian aspiration now lies in the hands of technological development. Unamuno in *Amor y pedagogía* (1995 [1902], 293–464.) asked "why do I want science if it does not make me happy?" Today we should be asking ourselves a similar question with regard to technology. The utmost satisfaction of mankind's desires through the growing technologization of life constitute the new guise taken on by the illusion of happiness, one that has become the dominant rule for education.

An alternative angle for guiding education would be the one put forward by Ortega y Gasset: we should educate "for a creative life" and not let ourselves be swept away by "practicism" (Ortega y Gasset 2004, II 406). A creative life cannot be reduced to mankind's adaptation to the environment: not everything is adaptation, despite the Darwinist attempts made to impose such a world-view (Ortega y Gasset 2004, II 406 and ff.; Nietzsche 1999a, b, 14 [123], 133, 303–305, 315–317). In Kant's wake, Nietzsche, Unamuno and Ortega y Gasset open up another sense of "adaptation", a creative and transformative sense; whereby, instead of focusing education on adaptation, as is proper to utilitarianism, they instead propose educating for the perfection of life, which also includes "some daring maladaptations". Education will have to be submitted to the imperative of vitality, which fosters the promotion – *züchten* ("rearing"), as Nietzsche would call it – (García-Granero 2017, 599–615) of vitally perfectible individuals. We should thus educate (nurture and not design) based on a creative (not mechanical) brain (Marina 2011; Amor 2015; Gracia-Calandín 2018a, b).

In order to educate in this way, Ortega y Gasset argues that education must start by nurturing the imagination, which in turn fosters the will to live. The creations of the imagination are comparable to a "psychic hormone" (Ortega y Gasset 2004, II, 419–421), which arouses enthusiasm and a vital sense of illusion, which are the best sentiments (as opposed to corrosive emotions) (Morgado 2017), and these provide a creative force that cannot be submitted to the criterion of utility.

Yet it is not easy to educate in the current information-rich and increasingly computerised environment (Floridi 2004; Marcos 2010), in which the new neuroscientific contributions (Cortina 2011, 2012a, b) are added to the growing technologization of human life. Thereby, we could interpret ourselves not only as informational organisms (*inforgs* and *cyborgs*) in a global environment (infosphere), but as *neuropoietici*, to the extent that the neurological dimension has been integrated into a configuration of the new image of mankind and its potential neuroeducation process.

During this era of technologization and the neurosciences, a critical moral philosophy (Ethics) has to be capable of discerning the possibilities and limits of the techno-sciences, and thus stress that the contemporary naturalization of intellectual discourse, now with its neurological guise, as well technologization, is the result of interpretations that proffer a new image of mankind and possibilities for education, for example those offered by neuroeducation.

1.4 Non-naturalistic Neuroethics for a Moral Neuroeducation

With the progress made in neuroscience the diverse facets of the foundations of human conduct are being revealed, and this is also the case for moral life, hence the relevance of neuroethics for moral education, for example, and with it proposals for bioenhancement, and even neuromoral enhancement,[8] based on tending to the brain through care and education. Following, the failure of traditional forms of moral education, and in response to the new opportunities offered by bioscientific and neuroscientific knowledge, in particular, there are some scholars who have proposed moral bioenhancement as an approach to a new moral education,[9] even including "neuroeducation".[10]

A desire to improve has been a driving force in education throughout the ages and new methods are currently being offered to enhance human biology, based on genetic and neuroscientific knowledge. However, this has nothing to do with any mere therapeutic use, but rather with measures for improving human potential, and thereby to extend the range of possibilities of human life through the use of biomedical technologies.

A fundamental problem is knowing what is meant by "enhancement", and specifically, "moral enhancement". The attitude of transhumanist movements (Diéguez 2016; López-Frías 2018), which as a general rule trust in the emergence of new possibilities for human nature without any clearly established moral criteria, do not suffice. Any possible moral enhancement requires a moral criterion to be established, it is not as simple as merely declaring: "become more of what you are", simply because until we know what that "more" consists of, there is scope for suspecting it may mean something "worse".

Indeed, one major issue faced by moral education is the question of the motivation for shaping and changing the attitudes of people and citizens. Everything we see in our world, day after day goes to show that traditional methods have seemingly failed in their task of educating. If we seek to be able to motivate people morally, "moral dispositions" will have to be improved, a task which will entail modifying the emotions, which are the basis of motivation, and that implies taking care of and educating the brain (neuroeducation). This opens up a new avenue for – at least – complementing education, because if we want to respond to the demands of a pluralist and democratic society, people's emotional systems will have to be addressed. A new possibility for doing this has seemingly arisen: treating the biological roots of moral motivation through drugs, implants and neural engineering, by means of which the most convenient emotions could be selected by genetic or

[8] See Cortina (2011).

[9] See Cortina (2012a, b).

[10] See Bettor et al. (2008); Marina (2011); Amor (2015); Gracia-Calandín (2018a, b), Codina (2015), Liao (2016), Sinnott-Armstrong (2016), Wiseman (2016).

neurological treatment (neuroeducation). But how do we know which emotions are most appropriate and useful in a moral sense?

As well as significant reports on this matter issued by international organizations, such as the OECD,[11] some thought-provoking and promising proposals have been set forth, such as that on the creative brain (the cultivation of neurological creativity or neurocreativity) elaborated by José Fuster y Marina (2015), María José Codina's neuroeducation based on cordial virtues (Codina 2015), the proactive epigenesis advocated by Kathinka Evers and Arleen Salles,[12] while Dacia Narvaez's (2016)[13] integrative ethical education is also worthy of consideration in spite of its risk of slipping back to a degree of naturalism.[14] But neither techniques, nor the neurosciences provide the moral content for moral education, and thus neither technoeducation, nor neuroeducation are in a position to replace education in general, and far less so moral education. What these proposals could indeed offer, nonetheless, is a better knowledge of the possibilities for changing certain capacities of the human being and the dynamism of human motivation. On the other hand, a wider knowledge of the possibilities these proposals can offer is not the same thing as having successfully used them to answer the question about the ethical reason for change (answering the "why?") and the moral significance of these supposedly improving changes.

Given the apparent failure of traditional education it is understandable that some educators should seek to take advantage of the new resources offered by neuroeducation, but according to Cortina, there are "limits" to what can be achieved through the proposal of moral enhancement through neuro-educational media. One basic limit is the need to obtain the voluntary consent of the subject to be submitted to an enhancement process, that is, to ensure there is moral autonomy, and for people to give their free consent they would first have to be morally motivated. It should also be asked whether there is a neurological medium capable of converting the feeling of respect for what is valuable in its own right, and not for the benefit it might provide, into a lived experience. Likewise, is there a technical or neurophysiological medium that can generate the ability to appreciate what is valuable for its own sake, so that the moral motivation caused by such an intervention on the emotions could replace the voluntary formation of character and thus provide a new real possibility for moral education?

Finally, one needs a firmly-established notion of the human subject in order to guide moral neuroeducation according to a non-naturalistic ethics. In my opinion this can be found in Zubiri's concept of the human being, insofar as this refers to an individual's constitutive and entitative structure in terms of the innovative notion of

[11] See Amor (2015); Gracia-Calandín (2018b, 201–215).

[12] See Pallarés-Domínguez (2016, 2019); Salles (2018).

[13] I would like to thank Javier Gracia-Calandín and Daniel Pallarés-Domínguez for sending some of the publications by this researcher, who attempts to show how to form one's own character: "who I should be" and "how am I achieving this?" (for example, by cultivating self-control and responsibility).

[14] See Gracia-Calandín (2018a).

substantiveness, an open essence and an appropriation of possibilities, which form – condition – the personal and moral life. A person's particular notion of sentient intelligence constitutes a decisive innovation amidst the dynamics of reality. Zubiri in fact interprets the new function of the human brain as a capacity for "hyperformalization", which constitutes the basis for opening up intelligence to the spheres of reality, freedom and human morality. Human reality as a substantive structure is above humanity's natural characteristics (*hyperkeímenon*), due to its open nature and the new properties achieved through appropriation.

The natural tendency of adapting to a medium is overcome through mankind's moral reality, thanks to the projective capacity of human intelligence, which requires a new sense of moral justification. A human being's capacity for his own adaptation is not determined by any natural properties, but it is through the hyperformalizing intelligence that a person opens up new vital and historical possibilities. Due to the new human capacity of hyperformalization, mankind takes charge of reality and is capable of enriching and preparing his life with new projects, values and historical creations. So, in spite of the primary function of intelligence being biological, its development moves in a transbiological direction towards freedom and a personal life, because the brain is perfectible and educable, even by incorporating the evaluative aspects of human life.[15]

José Antonio Marina reminds us that Luria stated that the human brain is not made to mechanically respond to stimuli (or in other words, for it to be causally directed from outside), but instead to anticipate plans for action (Marina 2011, 36–38; Fuster 2014; Cela and Ayala 2018). Julián Marías, drawing on the projective sense of the human being, as stressed by Ortega y Gasset, declared that we are cast towards the future, we are *futurized*; one should also remember that according to Ayala (1986) and Richart (2016), a sense of anticipation is an enabling trait of moral life. The important thing is that the human being projects, appraises and decides on the purpose of interpreting information. Education presupposes the capacity to perfect systems of interpretation, because within the human being there exists a broad space between the stimulus and the response within which one's own vital project is freely devised and chosen.

1.5 Humanistic Neuroeducation

Scientist, positivist and behaviourist schools of contemporary thought, ultimately governed by a naturalistic model, have sought to locate themselves "beyond" the notion of dignity.[16] The need to revitalise the humanistic and enlightened value of human dignity nevertheless continues to be a concern, and it structures the content

---

[15] See Kant (1989, 1992); Nietzsche (1978); Ortega y Gasset (2005, 2007); Adela Cortina (2009); Gracia (2017); Conill (2006).

[16] For example, Mosterín (2006, 379–385; 2014, 70–71, 77–79).

sourced from different traditions such as the biblical roots of education, because traditions such as these are a source of inspiration and life, from which whatever proves most valuable for a life of plenitude has to be reconsidered, selected and reinvented.[17]

Human dignity is the axiological key to modern humanism and in actual fact to transmodern humanism likewise. Despite the defence of animal rights, or of the intrinsic value of nature and life (Cortina 2009), the ethics that provides the moral foundation for the constitutions of many liberal democracies and the declarations of international institutions is the humanistic ethics of human dignity.[18] Both human rights and the arguments of applied ethics continue to have a common starting point in the concept of dignity as applied to the human person, and a non- naturalistic ethics can be sustained in this regard in order to act as a foundation for new proposals of moral neuroeducation, ones that refuse to be swept along by naturalization.

Education in general and neuroeducation in particular are based on a specific vision of human nature. The interpretation of human life, which takes into account the value of dignity, may be connected to the ethics of recognition: human beings not only have natural needs and desires, as do animals seeking self-preservation, but what they seek above all is other people's desire: they want to be recognised, which is the reason for their struggle for recognition. The target of their endeavours is not determined by mere biology, but by a wish for recognition, that is, by a new vital dynamism which could be interpreted as eleutheropathy (the pathos of freedom) (Nietzsche 1978, Section nine; Fukuyama 1992, 17, 216). Through innovative dynamism the brain is able to be educated in a free domain, and thereby overcome various social phobias and pathologies (Honneth 1994; Pereira 2013; Rosa 2013), such as aporophobia (Cortina 2017a, b); this is due to the fact that human beings, by virtue of their self-awareness and freedom "know they are above biological life" (Valls-Plana 1971, 119), and live in a new order of meaning, one proper to a personal life.

In this new order of human life, one has to learn to choose well. Our lives are fashioned through choices; we lead our lives choosing. Choice is an essential component of human life, one that is not only developed through working within the natural order, but also in the biographical order (Ortega y Gasset 2009, IX, 2019, X). Choice is voluntary, but not reducible to mere desire as appetite as it also involves rational activity, such as deliberation. Knowing how to choose forms part of an ethical virtue, choosing a good form of life, and the means most appropriate to carry out the purpose of a good life on the basis of feasible possibilities. The formation of character (*êthos*) takes place in this way, in response to the architectonics of the human being, shaped by desire and intelligence (thought): "desiderative reason" or "ratiocinative desire" (Aristotle 1999, V, 2). Neither logic nor a mathematical vision of the processes of rational choice suffices, nor does the mere calculation of probabilities; there is no algorithm for human rational decision-making.

[17] See Conill (2002, 2003).

[18] See, for example, Habermas 2011, 13–38; Cortina (1990, 2017a, b, 255–276).

A review of rationality is required to show that besides calculation there is also uncertainty, emotions, feelings, desires, aspirations, dreams and evaluations, each with their own neural basis, which have to be taken into account in order to focus education properly. For example, neuroeconomics, with its research into the decision-making processes, has been seen to involve not only a maximizing impulse, but also a sense of justice and cordial cooperation (Conill 2012, 2013; Damasio 2006, 2010; Morgado 2010; Cela and Ayala 2018; Calvo 2018).

The issue of choice leads us to consider different ways of understanding rationality, and now the neurosciences offers a further position to do so, and one of their main contributions consists in incorporating evaluations made at a neurophysiological level. Rationality is not axiologically neutral, since the act of evaluation is an activity that takes place in the most basic strata of the human being (Morgado 2010; Nietzsche 1978). It is already common practice to allude to the study of emotions in order to understand the process of choice and of decision-making: for example, evaluations entailing reward and punishment, as well as the vital values that regulate life in homeostasis processes and which guide conduct depending on survival and well-being. In short, decisions are "based on value". One chooses anticipating the value of the possible results. Everything is grasped with some "value indication" (Damasio 2006, 2010), with regard to the vital needs of survival and well-being. To live we have to assign value, taking into account the gains and losses, the rewards and punishments. Neuroscientific research can contribute a clearer knowledge of the "neural bases of reason" by incorporating its evaluative dynamism, from which the emotions and feelings emerge as forms of expression.

The education of the human individual has to involve these basic evaluations, ones which arise from the neurophysiological strata, and this is because life is not lived according to mechanical need, but by choosing and deciding within certain real (feasible) possibilities, which include preferring some options and discarding others in every lived circumstance (Ortega y Gasset 2010, X, 3–4, 158–163). Human life cannot be reduced to biological life in the psychosomatic order, as it has a biographical and creative sense, involving choice and preference. To choose well, however, one has to avoid whims (Ortega y Gasset 2010, X, 3–4, 158–163, 2009, IX, 583, 1172–74), since choice is what forms a personal reality through the appropriation of vital possibilities (Zubiri 1986; Aranguren 1958; Gracia 2007, 2017; Cortina 1990) and this is the way an individual becomes a moral reality: their moral condition depends on their ability to appropriate possibilities for themselves. "The moral sphere is demanded by the natural one", nevertheless; mankind can neither lack preferences nor desires. Nonetheless, what has been construed as "adaptation" in human beings must be understood as a form of appropriating certain possibilities while discarding others. Properties acquired in this way display a mode of being proper to mankind, one that is distinctive to this emerge naturally, since the former are the result of a chosen (preferred) appropriation of possibilities whose result is the configuration of a "moral reality" (Zubiri 1986, 344–345.).

Such an appropriation takes place by choosing from among possibilities, and it is based on an intellective function, which – even though this is bio-physiological – opens up a new field, due to the hyperformalizing structure of the human brain

(Zubiri 1986, 509–544, 516). Within this field of choice and appropriation a person gradually determines what he or she wants to become by means of the ongoing exercise of the ability to choose and prefer certain possibilities from among those offered over the course of a life. This is how personal and historical life is gradually created: by making vital choices and appropriating possibilities.

The possibility of moral education today involves neuroeducation: morally educating the person through a brain creating moral life (Marina 2011; Amor 2015; Cortina 2011; Calvo 2018; Codina 2015, Gracia-Calandín 2018a, b; Pallarés-Domínguez and Andrés Richart 2018), but it does not tend to tackle the issue of moral learning nor of moral education as such. At most one can find studies on reward and punishment, but not on the moral values involved in the process of people's education and which form their moral character from their earliest years. Nor are there many neuroscientific studies covering the moral education of the human being as an individual. Some significant studies could nevertheless be mentioned, whose main outlines I shall attempt to draw on below for a potential humanistic moral neuroeducation.

Those who wish to educate people must incorporate the study of neural mechanisms of learning, which also reveal "how experience changes our neurons". What needs to be known to further moral neuroeducation is whether this is also applicable in the case of moral experience: whether we are able to learn through the nervous and endocrine systems not only to solve problems of survival and well-being, but also to address issues regarding our post-conventional moral nature. Essentially, neural mechanisms enable us to choose and guide actions by means of education, incorporating the moral cultural mediations in which we live. We are capable of learning to give meaning to stimuli in order to construct a "personal matrix", which according to Marina consists of a "field of possibilities which gradually develops, by means of neural maturation, one's own actions and experience" (Marina 2011, 62).

Thanks to the plasticity of the brain this education process is possible, which is what enables the human being to invent and create. It is for this reason that the brain has to be cared for, it is central to education; neuroplasticity "lies at the source of the capacity to learn, to make use of culture, to be free" (Marina 2011, 80). The brain is not given to us fully formed, but what it can do must be discovered through its plasticity, which "enables us to be treated as a work of art" (Marina 2011, 85). Here reference to Nietzsche and Ortega is inevitable, the latter who considered the human being as a creator of himself, with the capacity to poeticise life (Nietzsche 1999a, b; Ortega y Gasset 2006, 272, 2009, 137, 352; Conill 1997). Neuroplasticity enables learning and self-education, drawing on experience and training, to thereby make "each brain [and each person become] a unique work". The possible care of the brain implies asking oneself which culture and which ethics is most appropriate for a moral neuroeducation, and furthermore, Marina proposes a "culture of neurological creativity", which requires the "adequate evaluation criteria" for the formation of character. It is essential to thus ask: adequate for what? In my view these criteria must be adequate for the humanisation of the natural world and the formation of an individual's character as a moral person: the brain has to be submitted to culture in order to develop an education based on a humanising ethics. And just as in other aspects of learning, it would be of great value if neuroscientific research were to reveal whether there are critical periods, or at least more sensitive ones, for undertaking moral neuroeducation, and thereby to "prepare a free and inventive brain that is able to [morally] advance our way of living and coexisting" (Marina 2011, 90). According to Marina, such progress is provided by a "universally valid project" within which everyone is the protagonist of their own life and attains their own moral autonomy. In such a project a prominent place is occupied by the virtues proper to human excellence, about which Marina believes there is a "consensus" between different cultures. I nevertheless think that this point would be worthy of more thorough philosophical reflection, instead of relying on a psychologized version of the virtues as "personal strengths", akin to Martin Seligman's discussion.

In any event, the inventive and executive capacity of human intelligence, given this has to be guided by certain "criteria for the evaluation of actions", requires ethics: "it is not a well-functioning brain that automatically produces good ideas, instead good ideas are those that enable us to educate a good brain" (Marina 2011, 182–183, 184). Education requires moral evaluation criteria; an ethics for educating the brain needs to be outlined, which is why the theory of intelligence proposed by Marina begins in neurology and ends with an ethics whose content is irreducible to neural processes. A "world of meanings" created through intelligence, including those of an ethical nature, emerges from the human brain and "feeds back" to the brain, without it being possible to reduce the significance, or the moral significance, to the neural domain. The aims of moral education are beyond scientific knowledge; they cannot be reduced to learning mechanisms, but the care of the brain must nevertheless be the purpose of a moral education, whose objectives are determined by an ethics that is humanistic and not naturalistic (Gracia-Calandín 2018a, b, 201–215). Such an ethics, one which overcomes reductionisms and is capable of guiding moral neuroeducation, is what has been proposed as an ethics of cordial reason (Cortina 2007; Codina 2015; Amor 2015; Calvo 2018; Gracia-Calandín 2018a, b).

References

Amor, José Ramón. 2015. *Bioética y neurociencias*. Institut Borja de Bioética, Esplugues.

Apel, Karl-Otto. 1973. *Transformation der Philosophie*. Frankfurt a.M.: Suhrkamp, Bd. II.

Aranguren, José Luis. 1958. *Ética*. Madrid: Revista de Occidente.

Aristotle. 1999. *Ética a Nicómaco*. Madrid: Centro de Estudios Políticos y Constitucionales.

Ayala, Francisco J. 1986. *Origen y evolución del hombre*. Madrid: Alianza.

Battro, Antonio M., Kurt W. Fisher and Pierre J. Léna (comps.). 2008. *The Educated Brain*, Pontifical Academy of Sciences and Cambridge University Press (*Cerebro educado*, Gedisa, Barcelona, 2016).

Calvo, Patrici. 2018. *The cordial economy – Ethics, recognition and reciprocity*. Cham: Springer.

Cela, Camilo J., and Francisco J. Ayala. 2018. *El cerebro moral*. Madrid: EMSE EDAPP.

Codina, María José. 2015. *Neuroeducación en virtudes cordiales. Cómo reconciliar lo que decimos con lo que hacemos*. Barcelona: Octaedro.

Conill, Jesús. 1991. *El enigma del animal fantástico*. Madrid: Tecnos.
———. 1997. *El poder de la mentira*. Madrid: Tecnos.
———. 2002–2003. La dignidad humana como concepto. EIDON 11 (October 2002- January).
———. 2006. *Ética hermenéutica*. Madrid: Tecnos.
———. 2012. Neuroeconomía y neuromarketing. ¿Más allá de la racionalidad maximizadora? In *Neurofilosofía práctica. Guía Comares de Neurofilosofía práctica*, ed. Adela Cortina, 39–64. Granada: Comares.
———. 2013. Neurorracionalidad y valor biológico. *Daimon* 59: 89–102.
———. 2015. Naturaleza humana en perspectiva biohermenéutica. *Pensamiento* 269: 1249–1260.
Conill, Jesús. 2016. La racionalidad práctica en perspectiva neuroética. *Pensamiento* 72 (273): 767–770.
Cortina, Adela. 1990. *Ética sin moral*. Madrid: Tecnos.
———. 2007. *Ética de la razón cordial*. Oviedo: Nobel.
———. 2009. *Las fronteras de la persona. El valor de los animales, la dignidad de los humanos*. Madrid: Taurus.
———. 2011. *Neuroética y neuropolítica. Sugerencias para la educación moral*. Madrid: Tecnos.
———, ed. 2012a. Neuromejora moral: ¿un camino prometedor ante el fracaso de la educación?, *Real Academia de Ciencias Morales y Políticas*. Madrid: Session of 8 January 2012.
———. 2012b. *Guía Comares de Neurofilosofía práctica*. Comares: Granada.
———. 2016. La conciencia moral desde una perspectiva neuroética. De Darwin a Kant. *Pensamiento* 72 (273), 771–788 y 827–848, respectivamente
———. 2017a. Eine diskursethische Begründung der Menschenrechte. In *Die Begründung der Menschenrechte*, ed. Margit Wasmaier-Sailer und Matthias Hoesch, 255–276. Mohr Siebeck.
———. 2017b. *Aporofobia. El rechazo del pobre*. Paidós: Barcelona.
Damasio, Antonio. 2006. *Error de Descartes*. Barcelona: Crítica.
———. 2010. *Y el cerebro creó al hombre*. Barcelona: Destino.
Diéguez, Antonio. 2016. *Transhumanismo*. Barcelona: Herder.
Flanagan, O. 2009. Varieties of naturalism. In *The Oxford handbook of religion and science*, ed. Philip Clayton, 430–452. Oxford: Oxford Handbooks Online: Oxford University Press.
Floridi, L. ed. 2004. *The Blackwell guide to the philosophy of computing and information*. Oxford: Blackwell Publishing.
Fukuyama, Francis. 1992. *El fin de la Historia y el último hombre*. Barcelona: Planeta.
Fuster, Joaquín. 2014. *Cerebro y libertad*. Barcelona: Ariel.
Fuster, Joaquín and Marina, José Antonio. 2015. El diálogo entre Neurociencia y Educación. *Participación Educativa* 4 (7): 8–10.
García Ruiz, Enrique, et al. 2018. *Ética y neurociencias. La naturalización de la filosofía*. México: Universidad Nacional Autónoma de Mexico.
García-Granero, Marina. 2017. Nietzsche y el mejoramiento humano. *Isegoría* (57): 599–615.
Goldman, A.I. 1995. Naturalistic epistemology. In *The Cambridge dictionary of philosophy*, ed. R. Audi, 518–519. New York: Cambridge University Press.
Gozálvez, Vicent. 2000. *Inteligencia moral*. Bilbao: Desclée de Brouwer.
Gracia, Diego. 2007. *Fundamentos de bioética*. Madrid: Triacastela.
———. 2017. *El poder de lo real*. Madrid: Triacastela.
Gracia-Calandín, Javier. 2018a. The non-naturalistic ethical end of neuroeducation. *Recercare* 22: 51–68.
———. 2018b. Ética para la neuroeducación. In *El desafío ético de la educación*, 201–215. Madrid: Dykinson.
Habermas, Jürgen. 2003a. *The future of human nature*. Cambridge: Polity Press.
———. 2003b. *Truth and justification*. Cambridge, MA: The MIT Press.
———. 2006. *Entre naturalismo y religión*. Barcelona: Paidós.
———. 2009. *Philosophische Texte*. Frankfurt a.M: Suhrkamp.
———. 2011. Das Konzept der Menschenwürde und die realistische Utopie der Menschenrechte. In *Zur Verfassung Europas*. Berlin: Suhrkamp.

Honneth, Axel. 1994. *Pathologien des Sozialen*. Frankfurt: Fischer.

Jesús Conill. 2019. *Intimidad corporal y persona humana. De Nietzsche a Ortega y Zubiri*. Tecnos: Madrid, 2019.

Kant, Immanuel. 1983. *Pedagogía*. Akal: Madrid.

———. 1989. *Metafísica de las costumbres*. Madrid: Tecnos.

———. 1992. *Fundamentación de la metafísica de las costumbres*. Madrid: Sociedad Económica Matritense.

Kohlberg, L. 1981. *The philosophy of moral development. Moral stages and the idea of justice*. New York: Harper and Row.

Liao, S. Matthew, ed. 2016. *Moral brains. The neuroscience of morality*. Oxford: Oxford University Press.

López Frías, Javier. 2018. The use of performance-enhancing technologies in sports through Nicolas Agar's "truly human enhancement" approach. *Performance Enhancement & Health*.

Machado, Antonio. 2004. Las Meditaciones del Quijote de José Ortega y Gasset [1915]. *Revista de Estudios Orteguianos* (8/9), 277–285

Marcos, A., 2010. Filosofía de la naturaleza humana. Eikasía. *Revista de Filosofía* (35), 181–208.

Marina, José Antonio. 2011. *El cerebro infantil: la gran oportunidad*. Barcelona: Ariel.

Morgado, Ignacio. 2010. *Emociones e inteligencia social*. Barcelona: Ariel.

———. 2017. *Emociones corrosivas*. Madrid: Ariel.

Mosterín, Jesús. 2006. *La naturaleza humana*. Madrid: Espasa-Calpe.

———. 2014. *El triunfo de la compasión*. Madrid: Alianza.

Müller, Axel. 2004. Pragmatismo y naturalismo eliminativista. In *El futuro de la filosofía*, ed. Francisco Galán and Ángel Xolocatzi, 80–96. México: Universidad Iberoamericana.

Narvaez, Dacia. 2016. *Embodied Morality*. New York: Palgrave Macmillan.

Nietzsche, Friedrich. 1978. *Más allá del bien y del mal*. Madrid, Alianza.

———. 1999a. Über Wahrheit und Lüge im aussermoralischen Sinne, KSA [Kritische Studienausgabe], G. Colli y M. Montinari (Hrsg.), Berlín, de Gruyter, Bd. 1, pp. 873–890.

———. 1999b. KSA 13 [Nachlaβ 1887-1889], 14 [123] and 133, pp. 303–305 and 315–317

Ortega y Gasset, José. 1983. *La razón histórica*. Madrid: Revista de Occidente/Alianza.

———. 2004. *Obras completas*. Madrid: Taurus, II.

———. 2004–2010. *Obras completas* [Volume IX (2009) y Volumen X (2010)]. Madrid: Taurus.

———. 2005. *Obras completas*. Madrid: Taurus, III.

———. 2006. *Obras completas*. Madrid: Taurus, «Aurora de la razón histórica», V.

———. 2007. *Obras completas*. Madrid: Taurus. VII.

———. 2009. *Obras completas*, Madrid: Taurus, IX: «La razón histórica».

———. 2010, *Obras completas*, Madrid: Taurus, X: «El hombre y la gente», pp. 175 and ff., particularly 178–181. Handwritten manuscript. Quote from p. 179.

———. 2016. ¿Naturalizar la idea de justicia? Una respuesta crítica desde la teoría moral de Jürgen Habermas. *Pensamiento* 72 (273): 827–848.

Pallarés-Domínguez, Daniel. 2016. Neuroeducación en diálogo: neuromitos en el proceso de enseñanza-aprendizaje y en la educación moral. *Pensamiento* 273: 941–958.

———. 2019. Moral neuroeducation: Proactive epigenesis and poverty. In *Moral Neuroeducation for a democratic and pluralistic society*, ed. Patrici Calvo and Javier Gracia-Calandín. Cham: Springer. [forthcoming].

Pallarés-Domínguez, Daniel and Andrés Richart, ed. 2018. Neuroethics and Neuroeducation: *Rethinking the Relationship between Neuroscience and Social Sciences*, Recerca. Revista de Pensament I Anàlisi (22).

Pereira, Gustavo. 2013. Elements of a critical theory of justice. Palgrave Macmillan.

Reports of the OECD. OECD/CERI 2007. Understanding the brain. The Birth of a learning science.

Richart, Andrés. 2016. El origen evolutivo de la agencia moral y sus implicaciones para la ética. *Pensamiento* (273): 849–864.

Rosa, Hartmut. 2013. *Beschleunigung und Entfremdung*. Berlin: Suhrkamp.

Salles, Arleen. 2018. Neurociencia y cambio moral: expectativas y límites. In Homenaje a Adela Cortina. Ética y Filosofía Política, ed Domingo García-Marzá, José Felix Lozano, Emilio Martínez y Juan Carlos Siurana, 395–405. Tecnos: Madrid.

Sinnott-Armstrong, Walter. 2016. The disunity of morality. In *Moral brains. The neuroscience of morality*, ed. S. Matthew Liao, 331–353. Oxford: Oxford University Press.

Sousa, David A. ed., 2011. *The Best of Corwin: Educational Neuroscience*. SAGE Publications Inc (*Neurociencia educativa*. Madrid: Narcea).

Unamuno, Miguel. 1995. Amor y pedagogía. In *Obras Completas I*, 293–464. Madrid: Biblioteca Castro/Turner.

Valls Plana, Ramón. 1971. *Del yo al nosotros*. Barcelona: Estela.

Wiseman, Harris. 2016. *The Myth of the Moral Brain.* The Limits of Moral *Enhancement*, Cambrige MA: The MIT Press.

Zubiri, Xavier. 1986. *Sobre el hombre*. Madrid: Alianza.

Chapter 2
Non-naturalistic Neuroethics for Moral Neuroeducation

Javier Gracia-Calandín

2.1 What Is the Future of the Learning Brain?

> After another 400 years of progress in technology and science, the biological basis of human intelligence has become enhanced. Electrical and chemical stimulation is improving brain function, and communications technology has merged with our nervous system, supporting the sharing of emotional experience across global digital networks. When famine strikes one part of the world, those on the other side of the planet feel the pain and suffering too, prompting immediate relief efforts and a global sense of hope. Every individual is a single component in a worldwide network of human experience. The human species is beginning to operate as a single community. We can prioritise broad, long-term goals of universal planetary significance, assuring survival of our species and the environment on which we depend. Moreover, again aided by technology, the species is editing its own genome, eliminating defects and improving our biological fit to the new world we are creating (Howard-Jones 2018: 176).

Paul Howard-Jones in his latest book, *Evolution of the Learning Brain*, proposes four potential outcomes for the human brain. In the third scenario he explores the possibility that the enhancement of human intelligence will be shaped by scientific and technological progress. Such a view is by no means new, but it merits closer scrutiny in order to consider the role of moral neuroeducation.

While the challenges of anticipating the future impact of science and technology are all too well known, there can be no doubt, that the realm of the possible can on occasions extend beyond the imagination and even science fiction. However, it should be acknowledged that already today a number of methods for improving intelligence have been developed and promoted, such as: brain enhancing drugs, Transcranial Electrical Stimulation (tES), genome editing, and internet connected neuronal implants. All of these offer plausible fields in which scientific and technological progress can contribute positively to the improvement of human intelli-

J. Gracia-Calandín (✉)
Universidad de Valencia, Valencia, Spain
e-mail: javier.gracia@uv.es

P. Calvo, J. Gracia-Calandín (eds.), *Moral Neuroeducation for a Democratic and Pluralistic Society*, https://doi.org/10.1007/978-3-030-22562-9_2

gence as well as to the common good of the human species. Yet, this is an optimistic position, given that scientific and technological progress is not always accompanied by moral progress, an issue evaded by Howard-Jones. Indeed, the only scepticism he expresses on this matter is that technological progress is not sufficiently well distributed among mankind. Yet this raises a key issue, which is that technology does not suffice in itself, and it needs to be applied correctly so it may contribute to the common good of mankind. The failure to address this issue is a recurring shortcoming of most approaches to neuroeducation or educational neuroscience, which either set aside moral questions, or simply take them for granted when discussing education.

In this regard, a fundamental issue that should be taken into account—a point made by Howard-Jones—is that technological evolution does not necessarily follow the path of what has been considered as moral progress over the course of history. It is, therefore, better not to blur technological evolution with concerns for progressive social development, and instead consideration should be given to the distinctive concepts of progress that have been developed along with their entwined moral tradition. For example, a distinction may be drawn between Darwin's theory of evolution and the eugenic theories formulated by Galton (Howard-Jones 2018: 11).

As is well known, the theory of evolution has been used to justify "scientific racism", most infamously during the Nazi era in Germany and the apartheid era in South Africa. However, like other seminal scientific ideas, the theory of evolution can be used for both good and evil; the crux of the issue is the ethical concept underpinning any theory's application. In some areas of the technology sector it may be argued that the problem of the immoral use of science and technology is simply solved by setting aside moral issues, but is this really the solution? Would it not be better to bring to the fore the underlying ethical model of education and justify why it should be foregrounded? Within the field of moral neuroeducation we argue that the solution is to explicitly address and promote an engagement with ethical concerns, indeed they are the foundation of neuroeducation.

2.2 The Need to Reflect on the Ethical Aim of Neuroeducation

Paul Howard-Jones in the passage quoted above and in his fourth and final scenario, besides distancing himself from naturalism through the distinction he draws between evolution and progress, also defends an ethical view of "global culture [as] capable of limiting conflict, climate change and loss of biodiversity". He argues that to achieve it, "education is probably the closest thing we have to ensure the well-being of human society" (Howard-Jones 2018: 180).

I wholeheartedly share the view that the key to moral enhancement is not achieved through technology, but education. However, in my opinion veiled allusions to ethics do not suffice. Instead, I deem it necessary to justify and demonstrate why ethics should be integrated into education and how neuroeducation can contribute to achieving this aim. Ethics is fundamental to education because it is embedded in the very concept of education (Gracia-Calandín 2018a).

There are different kinds of learning, some are morally beneficial, others are merely instrumental, while some are clearly morally harmful. What needs to be determined is the basis for deciding whether one kind of learning has led to the attainment of a moral good. Therefore, it is fundamental to consider the aims of education, which is a key issue for any kind of reflection on education, even when there may be different approaches to it (Marples 1999).

Richard Pring (2014) differentiates between a descriptive sense of education and an evaluative or normative sense. He identifies the descriptive sense with activities and institutions whose educational aim is limited to learning achievements that are demonstrated by merely stating that one studied at a certain school or university. In contrast, Pring's normative sense of learning addresses what it means to be an educated (or uneducated) person. Central to this normative sense is whether the student has become a better person in the broad sense of what is meant by a good person; considerations of whether someone studied at one university or another are not relevant.

When education is framed by an ethical approach it is not reduced to any consideration of the best means of making the system economically viable, but in line with the aforementioned normative and axiological sense of education, its overarching concern is to be valuable in its own right. Indeed, its ultimate goal is not to create a workforce better equipped to contribute to economic growth, but instead the formation of individuals in the full sense of a person's potential and skills. This second and genuinely ethical sense of education is forthrightly expressed in article 26 of the UDHR.

> Education shall be directed to the full development of the human personality and to the strengthening of respect for human rights and fundamental freedoms. It shall promote understanding, tolerance and friendship among all nations, racial or religious groups, and shall further the activities of the United Nations for the maintenance of peace (article 26, *Universal Declaration of Human Rights*, UNO 1948).

In line with this and more recently, the UNESCO report *Rethinking education. Towards a global common good?*

> The humanistic values that should be the foundations and purpose of education include: *respect for life and human dignity, equal rights and social justice, cultural and social diversity*, and a *sense of human solidarity and shared responsibility for our common future* (UNESCO 2015: 38).

Placing education within an ethical framework underscores how its specifically practical activity is not descriptive but instead, primarily normative, since it is concerned with regulating or guiding action with the specific goal of shaping a type of personality. Therefore, it makes sense to speak of good or bad education, firstly, because an approach based on an instrumental rationality of a strategic or mesological order will only attain the means best suited for a pre-established purpose, and secondly, because what is at issue is a consideration of what are the highest and worthiest goals to be aspired to when undertaking the thorough education of individuals.

Empirical research enables an understanding of biological elements such as the improvement of physical well-being or education for health (eating habits, psychometric development ...). It can also address aspects of a psychological and sociocultural nature, such as the sequential development of complex cognitive and emotional skills. Undoubtedly, the current knowledge of the cerebral basis of learning and teaching provided by neuroscience is very significant. However, empirical research is sterile and contributes little with regard to any degree of reflection on the aims and values that should guide and state what education should be, as opposed to merely detailing the facts about how it takes place. The limitations of empirical research in this regard are all too evident, as it is a further normative or evaluative level that articulates and permeates the definitive decisions taken regarding both the physical-biological and the psychological-social dimensions of education. What is at stake here are the variables required to determine what humanising improvements and development consist of.

It is primarily ethical research that has to be taken into account in order to develop a direct focus on values, habits and norms, as well as ideas about what is good and evil, that together govern the development of a human being's character and behaviour. It is ethics as moral philosophy that demonstrates the ultimate meanings of human development. If what is at stake is nothing less than defining what the "higher aims of education" should be, this can only be undertaken by defending a model of a fair society. Furthermore, a humanistic approach to education becomes fully meaningful within a model of liberal democracy in which education contributes to achieving rational autonomy (Hand and Davis 2016).

Neuroeducation is frequently considered apart from moral values and discussion to date has focused on its instrumental contributions, such as for calculation or reading comprehension. Needless to say, it is important to address these issues (Howard-Jones et al. 2016), but from the ethical point of view that underpins education, it is unjustifiably reductionist to exclude from neuroeducation any reflection on its guiding values, norms and principles.

Ethics enables reflection on issues in greater depth and with more rigour such as, for example, the controversial concept of "quality" when applied to education. In this case what is under scrutiny is not how specific goals are achieved, but rather the fundamental goals the education system seeks to achieve. Is the main aim of education, the supply of potential employees for the labour market? (Nussbaum 2010; Gracia-Calandín 2018d). From an ethical perspective it is clear that educational policies are based on, shaped and guided by certain "higher standards". Therefore, given that the values and quality of learning are being reduced to a quantification of knowledge that is defined by a variety of specific evaluation tests, it is a priority to undertake a reflection on what are education's ethical values and principles (Pring 2014; Mansell 2007). In fact, the search for empirical evidence upon which to base an educational model, as well as the assumption that where there is no measurable behaviour result there is no learning, are unequivocal indications of the spread of empirical approaches. Yet, is it not wholly inconsistent to base educational assessment on measurement parameters developed through empirical research (Smeyers 2010)? And, ought not educational evaluation, understood in its most radical sense, be something that ultimately gives rise to the values, principles and goals of education (Phillips 2005)?

When neuroscience is applied to education it contributes new explanatory models for understanding human learning. Some authors uphold the view that neuroscience is objective, neutral and disinterested. However, it should not be overlooked that wanting to know how the brain's learning processes function is a form of interest in its own right and, in my opinion, a very significant one. Yet this is not the only one, nor the principal one. In fact, scientific research is itself predicated on a range of ethical concerns such as, for example, an ethical commitment to truth within the scientific community. Therefore, to consider science and technology as guides to societies, while depriving them of any scope for reflection on the ethical framework that both guides their activity and assures the use of scientific and technical knowledge for the greater good is, to use Habermas's terms (1968), a mode of "ideology". One method of combating this issue is to address the fundamental question about the meaning and justification of scientific activity and findings. Thereby, it is necessary to ask whether, in light of the findings of neuroscience can it genuinely be understood as a normative scientific practice (Salles 2013)?

Bearing in mind the contemporary debate within philosophy and, above all, regarding ethics and neuroscience, in particular within the field of education, three entwined questions need to be addressed: should education modulate its aims, or even subordinate them to those of neuroscience? Should neuroscience naturalize education so that educational ethics become adapted to evolutionary patterns of behaviour? In short, what should the aim of neuroeducation be?

In their response to Bowser (2016), Howard-Jones et al. (2016) state that "educational neuroscience aims to motivate educational thinking and practice through models arising from neural and behavioural data" (Howard-Jones et al. 2016: 622). Through neuroimaging studies, new complex cognitive abilities are revealed that were not predictable solely on the basis of behavioural data. The example they provide is drawn from the learning of mathematics, and in particular children and adults' learning of integers and negative numbers. However, while I believe that the aim of seeking to develop instrumental skills (in the case of dyscalculia) or reading (in the case of dyslexia) is legitimate, it is nonetheless short-sighted to propose this as the ultimate and overarching goal of education. In contrast, in my opinion, it is fundamental to clarify the immanent ethical dimension of educational activity, above all, when consideration is given to the current rise in the use of naturalistic ethics to explain education. Should evolution set the tone and be the norm for moral education? Does the incorporation of neuroscientific findings into education practice inherently lead to the adoption of a naturalistic evolutionary ethic?

2.3 Neuroeducation in the Light of Non-naturalistic Neuroethics

A number of naturalistic approaches have become prevalent in the field of neuroethics, and these have sought to explain human behaviour on the basis of brain mechanisms, or the neurotransmitters involved in cerebral processes, as well as with key insights offered into organisms' adaptation to their environment by the theory of evolution. Such approaches argue that the brain's evolutionary heritage could provide explanations of both prosocial and antisocial behaviour depending on the context. However, such naturalistic explanations open up a horizon of explanation that requires careful scrutiny. The problem is one of reductionism, as naturalistic approaches such as these conflate the ethical view that should guide human behaviour with both the evolutionary purpose to which, according to the logic of environmental adaptation, the notion of a species' development or an individual or group's quest for survival. Once more a confusion between different senses of purpose arises. Should the distinction between ethics and evolutionary purpose be traced? This is a key question for moral neuroeducation, one which is fundamentally concerned with whether the moral realm has to be subordinated to the logic of biological evolution. The discussion that follows will address this issue.

Since Charles Darwin's *The Descent of Man* (1871) the aim of evolutionary approaches has been to trace the biological roots and phylogenesis of morality in order to better understand human behaviour (Cela and Ayala 2018: 51ss). According to Francisco Ayala (1987), the moral capacity of making choices and value judgements, as well as anticipating consequences, is the result of the eminent intellectual capacity of the human being, which is, in turn, the outcome of human evolution.

While, this phylogenetic approach has great potential to reveal the neurophysiological basis of the human species, according to Ayala (1987) and Cela et al. (2018), it would be harmful for ethics to confuse the human being's neurobiological constitution with the specific or particular content of a moral code. Were one to do so, normatively valid behaviour would be that which increases a capacity for adaptation, yet as is well known this is not always the case. In fact, a brief field study shows that although all human beings have the same neurobiological nature, very different moral standards are encountered within the diverse range of human societies. Ethical behaviour did not evolve because it contributed to human beings' adaptation, as many of the moral norms of advanced societies, such as monogamy, do not lead to an optimization of evolutionary performance; in regard to this example, polygamy would far more effectively foster evolutionary adaptation. Therefore, cultural evolution does not always go hand in hand with biological evolution, and to deny this would lead to the disastrous consequence of trying to replace ethics with socio-biology.

A fundamental distinction between moral norms and biologically determined behaviours should be kept in mind. Nonetheless, as Ayala points out, some moral norms can foster evolutionary optimization (taboo, incest, parental respect, the greater degrees of culpability attributed to female as opposed to male adultery…), while others are inconsistent with behaviour that favours natural selection (monogamy and non-polygamy by males, edicts to love all human beings). Therefore, the socio-biological argument that the only function of morality is the adaptation and conservation of intact genetic material is fallacious.

In pursuing this argument, maintaining a clear distinction between the cerebral basis for behaviour and its moral foundation is a key critical foundation (Cortina 2011: 94–96). While the neuronal basis alludes to a descriptive dimension, one that therefore falls within the competence of the neurosciences, the moral foundation alludes to a normative dimension, analysis of which is the task of moral philosophy. Clearly distinguishing between these two dimensions, and the disciplines trained to study them, avoids falling prey to both the naturalistic fallacy (when moving from one dimension to another) and naturalistic reductionism (reducing the latter dimension to the former). Regarding this foundational task, it should be pointed out that moral duty transcends the adaptive mechanism of evolution (Conill 2017; Gracia-Calandín 2018b).

With regard to the concern of this chapter, moral neuroeducation, it is of the utmost importance not to reduce the explanation of human behaviour to phylogenetic factors, and still more so, not attempt any derivation of the normative guidelines for education from these phylogenetic factors. In my view, Darcia Narvaez's (2016) theory of moral neuroeducation succumbs to this twofold failure; it advocates using the behaviour of our earliest ancestors, from 30 million years ago, as a frame of reference for ethical normality. Darcia Narvaez proposes that the "environment of evolutionary adaptation"of ancestral mammals determined the moral norms that were subsequently developed. It is not my intention to dispute the importance of tracing a neuronal basis for the ethics of care, a key facet of the legacy of mammalian evolution, nevertheless, a point is inevitably reached at which care itself, for reasons of the evolutionary group's own survival, evolves towards its limitations. Furthermore, Narvaez's theory is incapable of justifying why the aforementioned ethics of care should have been extended beyond the contexts of the family contexts and the species group. The naturalistic premise of guiding education according to environmental parameters of evolutionary adaptation can prove pernicious if what is proposed is a type of education based on universal principles such as the Universal Declaration of Human Rights.

Extending the scope of moral sensitivity and recovering a bond with nature is, in my view, the appropriate approach, one that would pose a vital challenge within highly technologized contemporary societies. Yet, an immense gulf divides this position from those advocating that "small communities of human hunters" fostered the "link with communal life" (Narvaez 2016: 16). Undoubtedly, care plays a central role in the early stages of the evolutionary development of child morality, but in ethical terms (and not only biological or social ones) such care must not be understood solely in an adaptive sense. If we do not broaden the ethical horizon to include a moral point of view in which it is possible to establish a hiatus between the value of nature and the value of human dignity, a child's moral development will, in my opinion, remain unnecessarily burdened. (Human) virtuous action does not attend to evolutionary factors, but to the cultivation of a moral character, and in regard to this neuroscience can make an important contribution (Marina 2011; Cortina 2011; Amor 2015; Codina 2015).

According to moral neuroeducation, it is principally education that makes human beings so extraordinary in comparison with other animals. We are not only what we are through biological evolution, but also through cultural transmission, and still more so through the education we receive. A phylogenetic perspective only offers an explanation of the cerebral basis of moral behaviour, and it should not be forgotten that among the most significant characteristics of our brain is neuroplasticity, which enables the brain to be shaped through culture and education. Education plays a decisive role in giving shape to both genetic expression and the brain's structures and functions. From the non-naturalistic ethical point of view there is nothing akin to determinism, except for the view that the human species is capable of directing its own evolution, thereby overcoming certain xenophobic or aporophobic tendencies of the brain. Moral neuroeducation studies precisely this evolutionary process, and not only at a biological level, but also in terms of moral improvement. In short, the fundamental concern of moral neuroeducation is the analysis of the relationship between the brain and education in order to achieve the formation of a good character (*êthos*).

2.4 The Contribution of the Moral Neuroeducation

2.4.1 Combating Demoralisation in Education

The famous OECD/CERI report (2007) *Understanding the brain. Birth of a learning sciences* highlights the fact that many students hate school and raises the question as to whether school is becoming hostile to the brain. Invoking the renowned words of Benjamin Franklin, the report highlights the importance of "involving" students in their education, as opposed to "teaching"them in the strict sense of the term, whereby education becomes true learning, as opposed to rote-memorization. To achieve this, a desire for knowledge is fundamental, as is succinctly summarised in the West African proverb, "Not to know is bad, but not to wish to know is worse". Therefore, it is appropriate to pause and consider what the neurosciences can say about learning motivation. The authors of the OECD / CERI report (2007) focus mainly on "intrinsic motivation":

> Motivation has a pivotal role in the success of learning, especially intrinsic motivation. The individual learns more easily if s/he is doing it for him/herself, with the desire to understand. Although it is currently difficult to construct educational approaches that could go beyond "carrot and stick" systems and target this intrinsic motivation, the benefits of this approach are such that it is of paramount importance for research to orient its efforts towards this domain (OCDE/CERI 2007: 27).

Neuroscience viewed from an evolutionary perspective can help understand and explain the importance of motivation as a link in the learning process. From a deep-time perspective (Howard-Jones 2018), "the engagement for learning" was originally found in the evolutionary roots of channelling our attention towards experiences that promised a form of reward. The human brain, over the course of its evolution as a vertebrate, has been gradually shaped by a reward system that activates certain fundamental parts of the brain such as the amygdala, the striatum or the accumbens nucleus in the processing of social and monetary reward (Izuma et al. 2008; Knutson et al. 2001), as well as through the deactivation of default mode networks by gamification (Howard-Jones et al. 2016). However, not only video games, but children's play more broadly is an immensely fruitful method for fostering learning (OECD/CERI 2007: 70–73).

Nonetheless, the human brain's reward system is not limited to this type of incentive, which in fact combines very significantly with factors relating to social reputation. The neural basis of reputation-based decision-making plays a fundamental role from childhood onwards (Izuma et al. 2012). More specifically in the area of education, social rewards such as praise, gold stars or points are fundamental and contribute to pupils' self-esteem and social standing, as well as to children's sense of novelty and curiosity (Howard-Jones 2010: 161). However, while the areas that are activated in the brain may be similar for various types of incentives, in my view, it is essential to recognize different modes of motivation with regard to their ethical significance.

In accordance with the distinction between intrinsic motivation and extrinsic motivation, I hold that an evolutionary explanation is necessary to better understand the conditions in which learning takes place, but it tends to illuminate just one facet of the issue, that of reputation and reward. However, consideration should also be paid to a type of intrinsic motivation directly linked to the learner's autonomy, one that depends less on reputation and more on moral principles and values, which are freely assumed by and directly linked to the learner's identity. With regard to this latter motivation it should be affirmed that there is a type of learning that is more directly linked to the increase of one's own competences, and less to the receipt of external rewards (Bain 2007: 45).

Ethical motivation is a challenge because essentially what is involved is the cultivation of character (*êthos*). Learners learn significantly when they link the new knowledge they acquire with their personal and social life (Horz 2012). Research shows that students who learned information in contexts drawn from their real life did so more efficiently than those who had to make inferences about contexts removed from their way of life (Reeves and Herrington 2009). The ethical goal for education is to cultivate a character that seeks to improve itself day after day, not just for the purpose of obtaining good qualifications, but instead for the gratification of performing the set tasks. Despite the effort that has to be made, curiosity is stimulated, a sense of interest prompted, and a specific competence is developed, as well as an understanding of why it is significant for the learner's life.

Furthermore, moral motivation is intrinsic and consists in challenging any possible manipulation that leads to action being guided by external reward and it seeks to foster in learners a "locus of control" for their behaviour (Bain 2007: 45). It should be noted that moral neuroeducation does not simply seek the application of self-control to particular tasks, but that self-control becomes personal autonomy when the behaviour of the learner is guided by moral principles that the learners establish for themselves; for example, the principle to respect all individuals as beings that have dignity and cannot be reduced to a price.

Once more, any lapse into an abstraction of impotent ideas must be avoided, and it is imperative to add that education focused on compassion is essential; yet, not compassion understood as concerning only the learner's social group, but in a full moral sense oriented towards the whole human race. As a result of neuroscientific studies, we are more concerned with the close-at-hand mundanities that happen to us, our families or friends, as opposed to distant tragedies of far greater magnitude. However, the model of moral neuroeducation presented here, opens up the possibility of cultivating compassion beyond our immediate sphere, one that encompasses every human being.

2.4.2 Eradicating Social Phobias

As has been explained by evolutionary theories, at least since Darwin (1871), the brain operates with mechanisms that have been developed through the process of evolution and these, although hidden, generate certain behavioural tendencies in people (Eagleman 2011). Like our evolutionary predecessors, we have an innate tendency to protect those who belong to our own social group, and also to reject strangers, whether they are foreigners or from a lower social stratum. An outcome of this are the behavioural traits of xenophobia (hatred of foreigners) and aporophobia (hatred of the poor) (Cortina 2017: 72ss).

From an evolutionary point of view, morality is indeed an adaptive mechanism, it provides a set of norms and values that enable us to survive, or in other words moral codes, which have underpinned the defence of social groups since the era of our hunter-gatherer ancestors (Churchland 2011). Essentially, these moral codes are closely linked to the emotions that were configured in the brain before the more advanced development of reasoning within the neocortex. The codes for caring for a social group and rejecting whosoever may represent a danger to one's own survival are very closely entwined with emotions, whereby in a concealed manner they still shape our lives as reflex and automatic responses, as has been demonstrated with renowned examples of personal and impersonal dilemmas (Greene 2014).

Nevertheless, from the point of view of moral neuroeducation the brain's evolutionary tendencies towards self-interest and concerns for control over our immediate environment, as well as protection against strangers and the rejection of the impoverished, do not have the final say; in contrast, it is possible to establish a pattern of moral conduct in accordance with the moral principle of a universal respect for human dignity. Furthermore, individuals share the suffering not only of members of their own social group, but also with those of another group. From this point of view, moral neuroeducation fosters compassion, a tendency that is universally present in the human species; what is more, a sense of compassion is inculcated, and this is not limited solely to members of a single group, but it extends in the form of care to all those in a disadvantaged or vulnerable situation. The moral dimension is made explicit because it is not simply a question of seeking the defence of the group itself, but of education strengthening the respect and freedom of each and every person. The empathic imagination and tendency for care crystallise upon the ethical horizon of compassion. Thereby, it is fundamental to consider the agent's point of view of and not just the observer's because it is only in this way possible to explain the qualitatively different motives and reasons for acting, as well as the normative character of moral habits, without both being reduced to mere social instincts.

Moral neuroeducation plays a key role in eradicating social phobias such as xenofobia or aporophobia because these are not only cerebral, more or less thoughtless tendencies. What is known as hate speech is based on a kind of deficient morality that must be overcome from the ethical point of view of a common and shared humanity. With regard to hate speech, people are not considered as singular individuals, but as collectives endowed with a trait that prompts rejection and contempt in the aggressor. The aggressor believes him or herself to be superior to the victim, thus leading to the latter's dehumanisation. However, for moral neuroeducation, it is fundamental that the full development of each individual's personality be always unique and with a greater polyphony than group stereotypes would allow for. Furthermore, it should be noted that moral neuroeducation has to combat statements regarding prejudicial facts that, are, nonetheless, difficult or even impossible to prove, and this is essentially, due to the fact that hate speech does not formulate arguments, but instead expresses contempt and seeks to disseminate it as widely as possible (Cortina 2017).

2.4.3 Cultivating the Values and Virtues of Civic Ethics Through Education

In certain positivist sectors, a type of teaching founded on facts rather than values has been promoted, on the basis that this type of education should prevail as it is objective and neutral. According to this positivist ideal, it is the empirically verifiable sciences that are the ones that should guide the learning process from beginning to end and all other questions must be avoided. Yet, has an education that solely fosters skills and instrumental knowledge, at the expense of the central role played by the capacities, attitudes and competences needed to generate both appropriate intellectual habits and the cultivation of character, not being wholly discredited?

With regard to the evaluation of education, the PISA and TIMSS reports, which claim to assess "academic performance" and "basic competences", in fact only focus on areas such as mathematics, science and, in the case of the PISA report, also reading comprehension. However, these reports raise the question: which social and civic competences constitute key basic competences (OECD 2003)? And it may thus be asked: what answers can be provided by moral neuroeducation? In my view, moral neuroeducation is not reducible to the positivist ideal of a mode of naturalism concerned with a devalued sense of education, one that has been converted into mere instruction. On the contrary, moral neuroeducation addresses the centrality of the world of moral values because people not only have the capacity to know about other beings, calculate numbers and speak foreign languages, they also have the ability to appraise and value, which is what enables us to make preferences, choices and orient ourselves as we lead our lives (Ortega 2005: 531–555).

Schools, along with other social agents, must also contribute to the education of how we assess and cultivate values, especially with regard to what are termed ethical-civic values and cordial virtues (Codina 2015). One of a number of important issues within the educational context is that opportunities are created for camaraderie and friendship to emerge. Friendship, as the seedbed for the broadest sense of civic friendship, is of the utmost importance and wholly necessary in democratic societies that practice and exercise deliberation as a powerful tool of civil society.

Fighting demoralisation through education, not only at the individual level, but also at the social level is, therefore, once again a way of giving credit to a values-based education and especially to the ethical values of active citizenship. These values can be summarized as significant freedom, a sense of responsibility, equality complemented by difference, active respect, solidarity with the most disadvantaged and, ultimately, justice in the pursuit of the full development of an individual's personality (Gracia-Calandín 2018a: 217–239). Being part of a democratic political community implies an idea of citizenship that entails ethical values that education must transmit at every opportunity and above all through school and family. If education ignores this ethical and civic dimension, then the democratic life of society will be seriously affected, thereby generating irresolvable social tensions and conflicts.

From the perspective of moral neuroeducation, as set out in this chapter, the main failure of school education is not that students do not attain the standards of knowledge established for their age (and this goal undoubtedly merits both improvement and encouragement). Instead, the most serious problem is that during the compulsory formal education stage, schools do not manage to develop students' personalities in an autonomous and competent way, whereby the students do not become compassionate people, who are both critical and self-critical towards injustice, and at the same time responsible citizens with a sense of solidarity. I consider this to be the main problem that must be addressed through an ethical education. Furthermore, while ethical education has to take centre stage in the classroom, its remit extends beyond the classroom to various spheres of civil society, and ideally to the realm of both the family and friends, as well as to other areas and associations of civic life (for example, civic solidarity organizations, the so-called NGOs).

2.4.4 *Rejecting all Forms of Moral Hypocrisy*

Moral neuroeducation is also opposed to moral hypocrisy, which is a form of schizophrenia of character, in which statements are dissociated from actions; a person preaches one idea and then acts in just the opposite way. Through moral neuroeducation this issue would be addressed by promoting a type of education that recognizes and draws on the decisive role that the amygdala and the entire limbic system plays in processes of teaching and learning, yet without restricting itself to this perspective. To achieve this attention must be paid to other factors related to how the neocortex affects reasoning. It should be noted, with regard to this, that different learning systems are directly interconnected.

In opposition to a type of neuroethics that reduces ethical-philosophical issues to a form of dualistic psychology, the moral neuroeducation proposed here does not view the brain's rational capacity in isolation from its emotional dimension (Gracia-Calandín 2018c). Thereby, rationality is not identified as a process of utilitarian calculation, but instead in terms of its connections to the affective dimensions of cerebral activity; because learning is also a way of configuring one's cultural, historical and corporeal life. It is through a language that attends to moral values and feelings that moral neuroeducation forges its impact, because it does not pretend to supplant the realm of life through empty idealisms, or by naturalistic scientific discourses that overlook the importance of genuinely moral values. The latter is one of the main shortcomings of educational neuroscience as it has recently and frequently been presented.

The neuroscientific explanation of moral hypocrisy can be located within the competition established between different parts of the brain that enter into conflict with another. Such conflict has often been explained as two processes of moral judgement, hot/cold, fast/slow, intuitive/calculated, rational/emotional. It is what is known as the dual system of cognition and has gained broad support in cognitive psychology (Kahneman 2011). Given that there is also a type of more reflective process I think it is important not to forget that even the most rational or calculating processes are activated by some type of emotion, because the brain is constitutively evaluative: it is the site of a struggle between different types of feelings that in turn give rise to more or less reflexive processes. Furthermore, reflection is effectively just one part of mental activity and a considerable amount of our mental processes occur at the level of non-reflective activities.

This explanation allows us to account for a mode of internal hypocrisy, for example one that consists in making statements about universal love and at the same time failing to recognize one's moral obligations to the rest of the world (Nussbaum 1996: 13). To avoid this first type of hypocrisy it would be advisable that education does not widen the abyss between declarations and actions in terms of the rejection of those who are distant from or different to us. Taking into account the brain's natural xenophobic and aporophobic tendencies, moral neuroeducation has to endeavour, through moral education, to counteract this tendency and foster the contrary response: the natural tendency of caring for others, to the point of extending care beyond the group itself, in order to nourish the bonds of compassion between human beings with the goal of achieving a sense of cosmopolitan hospitality (Cortina 2017).

In addition, in my view, another kind of internal hypocrisy can occur, according to which an individual may declare equal respect for every human being, while in practice favour only those at a distance from them, which would be to the detriment of one's immediate neighbours. This mode of hypocrisy can become the outcome of a type of abstract cosmopolitanism that generates individuals who are disconnected, and incapable of committing themselves to their immediate circumstance. It is akin to escapism, and individuals sympathize from afar, while ignoring those closest to them (Bok 1996).

To counter this second type of internal hypocrisy, based on the aforementioned abstract and uprooted cosmopolitanism, moral neuroeducation foregrounds the central role of emotions and feelings for the formulation of moral judgements and more broadly in moral life as a whole. In this regard, consideration must be given to how moral judgements have to be accompanied and motivated by the cultivation of emotions that underpin a person's development of a sense of compassion for whoever suffers (both from their immediate social group itself or from another): this is the core of all authentic ethical reasoning.

Moral neuroeducation is opposed to the theoretical teaching of morality based on abstract statements that have no roots in life and that end up generating a form of dualism. Moral neuroeducation therefore affects the formation and cultivation of character through an eminently practical teaching. This does not imply that it is subject to particular interests, instead its moral dictates are rooted in daily practice. It consists of an education founded on the world of emotions, feelings and values, and upon this cultivating realizable ideals of reason. If this is not pursued as has been discussed there is the risk of morally reprehensible phobias that can give rise to hate speech. Faced with such issues, the realizable ideals of practical reason enable the cultivation and nurturing of feelings through an embodied rationality. Furthermore, it is precisely through such feelings that the ethical-civic language that we employ gains all its vital and experiential significance. It is a language that above all expresses personal experience and our shared coexistence in democratic, egalitarian and pluralistic societies.

2.5 Conclusion

Moral neuroeducation must be based on a knowledge of brain mechanisms in order to identify cognitive and affective processes that are the basis of certain types of behaviour. However, this does not imply denying or failing to sufficiently recognize the space itself of the world of values and moral standards. Furthermore, if possible, these values and standards should guide the educational process, which is why it is so important not to confuse neuroscientific experiments using neuroimaging with the educational activity of teachers. The model of moral neuroeducation advocated in this chapter is informed by neuroscientific research, yet without disregarding the non-naturalistic ethical goal of education. Indeed, the position advocated here is that education should forge a critical, compassionate, responsible, and cosmopolitan citizenship that encompasses a strong sense of solidarity. Working from this non-naturalistic ethical approach, moral neuroeducation is able to take advantage of all the neuroscientific discoveries that trace how the brain functions and use them to make an important ethical contribution to society. Among the major contributions of moral neuroeducation set out here are the ability to combat demoralisation in education, the eradication of hate speech, the cultivation of the values of a civic ethics and the elimination of every kind of moral hypocrisy.

References

Amor, José Ramón. 2015. *Bioética y neurociencias*. Esplugues: Institut Borja de Bioética.

Ayala, Francisco. 1987. The biological roots of morality. *Biology and Philosophy* 2: 235–252.

Bain, Keneth. 2007. *Lo que hacen los mejores profesores de universidad*. Valencia: PUV.

Bok, Sissela. 1996. From part to whole. In *For love of country?* ed. Martha C. Nussbaum and Joshua Cohen, 38–45. Boston: Beacon Press.

Bowser, Jeffrey S. 2016. The practical and principled problems with educational neuroscience. *Psychological Review* 123: 600–612.

Cela Conde, J. Camilo, and Francisco J. Ayala. 2018. *El cerebro moral. Evolución del cerebro y valores humanos*. EMSE EDAPP.

Churchland, Patricia. 2011. *Braintrust. What neuroscience tell us about morality*. Princeton: Princeton University Press.

Codina, María José. 2015. *Neuroeducación en virtudes cordiales. Cómo reconciliar lo que decimos con lo que hacemos*. Barcelona: Octaedro.

Conill, Jesús. 2017. ¿Tiene arraigo en el cerebro la libertad? *Pensamiento* 73 (276): 493–514.

Cortina, Adela. 2011. *Neuroética y neuropolítica. Sugerencias para la educación moral*. Madrid: Tecnos.

———. 2017. *Aporofobia, el rechazo al pobre. Un desafío para la democracia*. Barcelona: Paidós.

Darwin, Charles. 1871. *The descent of man, and selection in relation sex*. London: John Murray.

Eagleman. 2011. *Incognito. The secret lives of the brain*. Edimburgo: Canongate.

Gracia-Calandín, Javier. 2018a. *El desafío ético de la educación*. Madrid: Dykinson.

———. 2018b. Crítica a la naturalización del deontologismo en la teoría del proceso dual del juicio moral de Joshua Greene. *Isegoría* 58: 205–219.

———. 2018c. Propuesta de neuroética no naturalista. In *Juicio moral y democracia. Retos de la ética y de la filosofía política*, ed. A. Richart, M. García-Granero, C. Ortega, and L. De Tienda. Granada: Comares.

———. 2018d. El fin ético no naturalista de la neuroeducación. *Recerca. Revista de Pensament i Anàlisi* 22: 51–68.

Greene, Joshua. 2014. Beyond-point-and-shoot morality: Why cognitive (neuro)science matters for ethics. *Ethics* 124 (4): 695–726.

Habermas, Jürgen. 1968. *Technik und Wissenschaft als «Ideologie»*. Frankfurt a.M.: Suhrkamp.

Hand, M., and R. Davis. 2016. *Education, ethics and experience. Essays in honour of Richard Pring*. London/New York: Routledge.

Horz. 2012. Situated prompts in authentic learning environments. In *Encyclopedia of learning*, 3086–3088. New York: Springer.

Howard-Jones, Paul. 2010. *Introducing neuroeducational research. Neuroscience, education and the brain from context to practice*. London/New York: Routledge.

———. 2018. *Evolution of learning brain. Or how you got to be so smart*. London/New York: Routledge.

Howard-jones, Paul, D. Ansari, Bert De Smet, D. Laurillard, S. Varma, B. Butterworh, U. Goswami, and Michael S.C. Thomas. 2016. The principles and practices of educational neuroscience: Comment on bowers. *Psychological Review* 123 (5): 620–627.

Izuma, Keise, Daisuke N. Saito, and Norihiro Sadato. 2008. Processing of social and monetary rewards in the human striatum. *Neuron* 58 (2): 284–294.

Izuma, Keise. 2012. The social neuroscience of reputation. *Neuroscience Research* 72 (4): 283–288.

Kahneman, D. 2011. *Thinking, fast and slow*. New York: Farrar, Straus and Giroux.
Mansell, W. 2007. *Education by numbers: The tyranny of testing*. London: Politico's Publishing.
Marina, José Antonio. 2011. *El cerebro infantil: la gran oportunidad*. Barcelona: Ariel.
Marples, R. 1999. *The aims of education*. London/New York: Routledge.
Narvaez, Darcia. 2016. *Embodied morality. Protectionism, engagement and imagination*. London: Palgrave Macmillan.
Nussbaum, Martha C. 1996. Patriotism and cosmopolitianism. In *For love of country?* ed. Martha C. Nussbaum and Joshua Cohen, 3–20. Boston: Beacon Press.
———. 2010. *Not for profit. Why democracy needs humanities*. Princeton: Princeton University Press.
OECD 2003. *Definición y Selección de Competencias clave (DeSeCo)*. Recuperado de http://goo.gl/ElpuCX (25/02/2015).
OECD/CERI. 2007. *Understanding the brain. The birth of a learning science*. Paris: OECD.
Ortega y Gasset, José. 2005. Misión de la Universidad. In *Obras completas*, vol. IV. Madrid: Fundación José Ortega y Gasset.
Phillips, D.C. 2005. The contested mature of empirical educational research (and why philsophy of education offers little help). *Journal of Philosophy of Education* 39 (4): 577–597.
Pring, Richard. 2014. From disguised nonsense to patent nonsense: Thinking philosophically. *Revista Española de pedagogía* 258 (72): 249–261.
Reeves, T.C., and J. Herrington. 2009. Authentic tasks: The key to harnessing the drive to learn in memebers of 'Generation Me'. In *Looking toward the future of technology enhanced education: Ubiquitous learning and the digital native*, 205–220. IGI Global. Disponibletambiénen https://www.researchgate.net/publication/267690756 (consultado el 1 de septiembre de 2017).
Salles, Arleen. 2013. On the normative implications of social neuroscience. *Recerca. Revista de pensament i anàlisi*. 13: 29–42.
Smeyers, P. 2010. Statistic and inference to the best explanation: Living without complexity. In *Educational research: The ethics and aesthetics of statistic*, ed. P. Smeyers and M. Depaepe, 161–176. New York: Springer.
UNESCO. 2015. Rethinking education. In *Towards a global common good?* Paris: Unesco.

Chapter 3
Moral Neuroeducation from a Phylogenetic, Ontogenetic and Functional Perspective

Andrés Richart

3.1 Introduction

Education performs a central role in guiding the brain's epigenetic development (Evers 2010; Pallarés-Domínguez and Richart 2018). Education is a complex process, through which knowledge, customs, values and forms of behaviour are transmitted among individual members of a society or group. The traits acquired through education are historical, cultural and contingent, and are adopted in a specific form by each individual (Ayala and Cela Conde 2001). Their cultural content requires a psychobiological structure as a foundation, which provides a neuronal support that renders the subject susceptible to being educated (Aranguren 1994; Hauser 2008; Cortina 2012, 2017). The brain is that foundation and it establishes the neurological conditions of the possibility of education.

Education can be thus be understood according to two strata: structure and content. Structure refers to the physiological basis underpinning the education process, which, existing from birth, undergoes a continuous development in response to an individual's environment. Therefore, moral education may be defined in a neuroscientific sense as a process that guides the epigenetic development of moral structures during ontogenesis, and it thereby enables the acquisition of cultural features, here referring to the content strata, which include the *ethos* or moral character, as well as a number of social and moral competences and skills. In order to study the natural conditions of the possibility of morality it is necessary to know its compositional elements, its origin, its development processes and how it functions. The aim of this study is to shed light on these issues and so enable the development of models and tools that may contribute to a fruitful moral education.

A. Richart (✉)
Universidad de Valencia, Valencia, Spain
e-mail: Andres.Richart@uv.es

P. Calvo, J. Gracia-Calandín (eds.), *Moral Neuroeducation for a Democratic and Pluralistic Society*, https://doi.org/10.1007/978-3-030-22562-9_3

3.2 Phylogenesis, Ontogenesis and the Functionality of Moral Agency

A key contribution to a deeper understanding of the subject under discussion is provided by Ayala's book *La evolución de un evolucionista* (1987), in which he sought to specify the necessary and sufficient conditions required for a human being to behave in an ethical manner. Ayala sets out these possible conditions in relation to a consideration of three abilities a moral agent must be capable of, and how a degree of psychobiological capacity is involved in making these abilities to possible. Therefore, drawing on Ayala's proposal a number of conclusions may be drawn about the emotional and cognitive elements that enable any human being to act as a moral agent.

The possible conditions that Ayala identified are as follows: an ability to anticipate the consequences of one's actions, an ability to make value judgements and an ability to choose between alternative courses of action. In other words, a moral agent must be able to imagine different possibilities with regard to the different possible courses of action open to them, and the various consequences that could result from them. In addition, the ability to make value judgements would guide an agent's decision-making when faced with the choice between acting in one way or another. Any decision will only be genuinely moral if the agent has free will during the decision-making process, whereby what is required is the ability to choose between various possible courses of action. Darwin, in *The Origin of Man*, describes the moral agent in a very similar way. He considers a moral agent to be characterised by the ability to represent past and future actions and motives, and also by the ability to form moral judgements, which provide a foundation for the agent's motives and actions to be approved or rejected (Darwin 1979: 107). Darwin did not consider the issue of free will, which is fundamental to Ayala's third condition, the ability to choose between various course of action, and it may be supposed that Darwin considered it to be necessary, since he implies that it is a prerequisite for the deliberation on and approval of an action or motive as moral.

The three conditions identified by Ayala, as well as Darwin, signal the existence of various psychological faculties. The ability to anticipate one's own actions requires the use of imagination, memory and abstract thinking, which would also permit a projection of the present situation as a series of possible hypothetical, future scenarios. The ability to form moral judgements consists of making judgements about our reasons for acting in one way and not another. Therefore, the ability to make value judgements is the result of metacognition, which is in turn associated with different faculties, like reason, language and abstract thinking, all of which have been developed to a considerable degree of complexity. In addition to these are other mental faculties, such as memory and imagination, and finally, we must also add prosociality. For some time research in the fields of biology and psychology has signalled an issue that neuroscience has today confirmed in detail: the exercise of reason within moral judgement incorporates emotional and prosocial factors that condition how judgements are made, and that guide the judgement-making process in certain directions (Moll 2002a, b; Greene 2003; Rizzolatti and Sinigaglia 2006; Morgado 2010; Damasio 2005, 2008; Haidt 2001; Churchland 2013; Salles 2015). Therefore, just as Darwin pointed out, human prosociality is necessary for the making rational moral judgements, in regard to which the concept of cordial reason is relevant here, as is demonstrated below (Cortina 2009). Essentially, a rational being needs prosociality to become a moral agent.

For millions of years emotions have mediated between all species' perceptions of the environment and the responses to it provided by their animal organisms (Damasio 2008). In many species, as the group became the main environmental factor that individuals needed to adapt to, their emotions evolved so as to offer appropriate responses to this environmental factor (Colmenares 2005; Morgado 2010). The acquisition of reason and abstract thinking occurred later, and these contributed a complementary function to that of prosocial emotions and the other faculties, yet without replacing the role played by the latter (Goldberg 2002; Forbes and Grafman 2010).

A key question that merits careful scrutiny is whether human beings comply with Ayala's three conditions. Firstly, consideration will be given to an individual's ability to anticipate future actions. Following that attention will be devoted to the issue of free will, and then, finally, an in-depth analysis of the ability to make evaluative judgements is undertaken. With regard to the anticipation of future actions, as has already been indicated, this is an ability that has been identified in human beings as well as other animal species.

Regarding free will, from a naturalist perspective there is no evidence to establish whether the human sense of free will is merely the result of conditioning, or is in fact determined by human biology; neuroscience is currently investigating this issue (Libet 1999; Roskies 2010a, b; Ortiz de Landázuri 2016; Pallarés-Domínguez 2013, 2016). While free will is a perquisite for moral agency, we can only make the assumption that human beings are really free in this sense. In any case, if free will does exist, it would require complex cognitive abilities, and the human being is one of the animals that is most likely to have achieved the necessary cognitive complexity that would make free will possible, albeit hypothetically speaking.

Turning to the issue of the ability to make moral judgements, it is essential that a social being acquired this ability in order to become a moral agent. A number of human faculties, such as reason, which can also be spatial and instrumental, language and abstract thought, when combined with other faculties, such as memory and imagination, enable the emergence of complex forms of cognition; a number of these faculties have also been identified in other animal species, as well as our ancestors. Metacognition and self-consciousness develop when various combinations of these faculties, especially reason, language and abstract thought, embrace greater levels of complexity. In conjunction with this, prosocial faculties and emotions, based on the assumption that free will exists, constitute a concurrence of elements that enable the emergence of moral agency. In other words, the concurrence of these elements may be identified as the natural conditions underpinning the possibility of moral agency.

In addition to the three aforementioned conditions discussed by Ayala and preempted by Darwin, a fourth condition must now be added, which is a capacity for mutual recognition. Although for both Ayala and Darwin the idea of recognition figures implicitly as a prerequisite condition for morality, it is necessary to explicitly state why it is needed. Social faculties and emotions highlight the need for recognition between social animals, this is still more relevant for humans, as they are moral agents (Cortina 2009, 2017; García-Marzá 1992; Habermas 2006; Apel 1985; Honneth 1997).

Ayala also states that morality does not develop in humankind as the result of evolutionary adaptation, which Darwin himself intuited, and this view is also held by authors such as Damasio, Nichols and Prinz, all of whom have argued that the origin of moral agency was has nothing to do with it favouring natural selection (Damasio 2005: 160; Prinz 2011, Nichols 2005). Instead, it may be stated that it originated as an evolutionary by-product and consequence of the acquisition and concurrence of certain psychological faculties (Richart 2016).

3.3 Moral Agency Studies as Foundation for Reflection on Key Issues in Moral Neuroeducation

The study of the natural conditions underpinning the possibility of morality reveals the entwined relationship between reason, language, abstract thinking and prosociality, as well as the roles played by the social emotions and faculties. Research in this area also reveals the possibility of modifying the brain, to the extent that it is malleable. Taking these two issues as a point of departure a series of guidelines for moral neuroeducation may be formulated. Over the course of the following sections attention is devoted to three key areas. The first concerns the possibility of corrupting the brain during epigenesis due to its plasticity. The second focuses on prosociality and human biological predispositions. Finally, the third area, addresses the concept of cordial reason, which is of particular interest because it integrates the rational and emotional dimensions of thought within human moral agents.

3.3.1 Brain Plasticity and Proactive Epigenesis

Brain structures are subject to continuous change due to the process of neuronal development, known as epigenesis, through which individuals develop in accordance with their interaction with their environment. According to Evers, with regard to the Changeux model, epigenesis refers to the selection of brain synapses being established according to their activity, which is a form of cultural imprinting (Evers 2010: 137; Salles 2017a, b). Brain development reveals an impressive versatility; 70% of the brain develops epigenetically (Cortina 2012: 220). Neural or synaptic plasticity, also called neuroplasticity, is a property of neuronal functioning that is related to the process of individual neuron connections and their communication with other neurons. Each neuron forms about 50,000 connections on average, which reveals the striking extent of our brain's plasticity (Evers 2010: 138; Jácomo and Serrão 2013).

The findings on brain plasticity are especially relevant for moral education. The neural plasticity and epigenesis that the human brain undergoes signals the possibility of how education can guide the development of brain structures, and this includes the moral dimensions of thought too. An individual's brain structure undergoes a distinctive configuration depending on how environmental factors are presented, whereby both educators and society have the possibility and responsibility to undertake both education and socialization in such a manner that they shape brain structures in the most appropriate ways.

Evers has developed a valuable proposal in response to this issue, one that explores the possibility of being proactive from an epigenetic perspective; given that, to a great extent, brain structures are shaped and developed through interaction with the environment an active role can be taken in constructing such cerebral structures. As a result, neural structures can be produced in individuals' brains that conform to the values and norms of their specific society or culture (Evers 2010: 136–144; Evers 2015). Thereby, the genetic predisposition conditioning moral and social behaviour in a selfish manner (Waal, Tomasello, Ayala) can be compensated through a proactive epigenesis. Through education we can interweave the structural element of morality with specific moral contents (Evers 2010: 136).

In addition, and drawing on recent genetic studies, Evers raises the possibility that certain epigenetically acquired brain characteristics could result in a genetic imprint that could in turn be inherited. In terms of moral education this data is especially relevant, as undertaking the education of values such as peace or cordiality could lead to the generation of a genetic imprint that would in turn give rise this type of moral value becoming innate for future generations. The same would happen were a person to be educated in values that we consider reprehensible, such as violence and hatred (Evers 2015). Therefore, with regard to the idea of a proactive epigenesis, educators and society as a whole must accept their moral responsibility (Salles 2017a, b; Evers 2010). At a genetic, biological and neuronal level, individuals may present predispositions that facilitate or hinder the acquisition of certain skills or a certain moral character, nevertheless neuronal plasticity offers the possibility of addressing any such predispositions (Jácomo and Serrão 2013). The fundamental role of education is once more underscored.

3.3.2 Prosociality and Biological Predispositions

Through evolution the human has become a complex animal, and this is above all noted with regard to the social and moral dimension of mankind. On the one hand, codes of social behaviour are embedded in our biology, whereby they predispose us to care for those closest to us, while also providing a biological basis for selfishly oriented forms of social and moral behaviour (Waal 1997, 2007; Tomasello and

Vaish 2013; Ayala 1987; Ayala and Cela Conde 2001; Greene 2013; Churchland 2013; Evers 2010; Cortina 2009, 2012, 2017). On the other hand, some moral values and norms are culturally acquired, which in many societies are claimed to be universal (Habermas 2010; Korsgaard 2000; Cortina 2007, 2009). As a result, a dissonance is generated between what biology defines and what ethics proposes. This fact poses a challenge in many senses, one that education must also face. As far as education is concerned, it must be asked what roles are to be granted, on the one hand, to the innate evolved biological codes and, on the other, to the universalist cultural codes (Cortina 2012: 221–222; Cortina 2017).

To avoid logical errors that lead to any form of naturalistic reductionism, we should resist considering the existence of aspects of the innate biological codes as a criterion or reason for formulating moral prescriptions. A rigorous knowledge of these biological codes undoubtedly provides valuable data that must be taken into account when delivering a moral education (Moore 1983; Teehan and Dicarlo 2004), yet they alone cannot provide a foundation for any proposal concerning moral duty. Instead, any ethical foundation must be developed from the perspective of moral philosophy (Cortina 2012; Habermas 2006, 2010; Apel 1985; Korsgaard 2000; Haidt 2001). In addition, Cortina has highlighted the issue of the natural sciences' inability to address the issue of the ethical foundation within their naturalist paradigm (Cortina 2007). Moral education, therefore, must address both moral philosophy and science, but without confusing the competences of each (Cortina 2012).

Since human beings have a tendency to act selfishly embedded in their nature, the task of moral education should be aimed at epigenetically modifying this human predisposition. Admittedly it is not possible to fully achieve, nor should it be an aim that individuals divest themselves of their biological, emotional and innate characteristics. The focus should rather be on training individuals to be capable, firstly, of recognizing these inclinations and preferences as innate biological tendencies, and, secondly, of understanding that they do not conform to moral arguments; in short the goal should be that people are able to be critical of such biological codes and reject them when there is no rational argument for them (Haidt 2001; Cortina 2012). In this way, moral education can form moral agents capable of reconciling their biological and emotional tendencies with a rationally based ethical position (Evers 2010: 136–144), and furthermore by engaging with the cordial facets of the exercise of reason (Cortina 2009, 2017).

3.3.3 Moral Neuroeducation and Cordial Reason: An Emotive-Rational Moral Education

The neurosciences have highlighted the important role played by the emotions in a number of brain processes and psychological faculties (Salles 2015; Damasio 2008; Einsengberg 2000; Greene 2003; Haidt 2001, 2013; Le Doux 1999; Morgado 2010; Churchland 2013). Ever since the earliest development of human emotions, their role has been to process the significance of perceived elements of the environment and offer an appropriate response to them. The development of the neocortex, which introduced complex forms of rational, abstract and metacognitive thought, occurred relatively recently in human evolution, a few hundred thousand years ago. Therefore, brain processes enabled by the neocortex often complement the cerebral role of the emotions, yet without replacing the latter (Damasio 2008; Ayala 1987; Ayala and Cela Conde 2001; Churchland 2013; Darwin 1979; Forbes and Grafman 2010; Goldberg 2002; Greene 2003; Haidt 2001, 2013; Tomasello and Vaish 2013; Waal 1997, 2007). However, human reasoning, which is associated with neocortex development, is in some cases able to transform, govern, revise, modify and substitute emotions.

So, it may be argued that neuroeducation must address human beings' emotive-rational character, and in particular explore the possibility of educating the emotions. The education of the emotions must be exploited to resolve the friction between biologically defined behavioural tendencies, on the one hand, and ethical conventions, on the other (Evers 2010: 125–133; Cortina 2009, 2012, 2017; Salles 2015). To successfully achieve this, a two-pronged approach must be taken. Firstly, emotions must be educated in such a way that a moral feeling become universally pervasive, even though it may be an innate emotion. Secondly, reason should be educated in such way that emotional intuitions about what is right and wrong may be subjected to an appropriate critique, by reviewing, correcting and discarding them as necessary (Haidt 2001, 2013; Cortina 2007, 2009, 2012; González-Esteban 2013; Reverter-Bañón and Medina-Vicent 2018; Cayuela 2012). Once this is done, the recipient of such an education will become a critically aware and morally responsible agent, one whose moral dimension engages with the emotions, but also subjects them to rational supervision.

3.4 Conclusions

Over the course of this chapter, the study of the natural conditions of morality, as undertaken within neuroscience and their sister disciplines, has been addressed from phylogenetic, ontogenetic and functional perspectives. A range of key elements that shaped the emergence of morality have been considered, which have in turn provided invaluable data for education. As has been discussed, there are some innate biological predispositions that condition social and moral behaviour in a selfish manner. However, most brain development occurs through interaction with the environment, and this, on the one hand, provides an opportunity to reinforce those tendencies that are considered valuable, for example those that favour prosociality. On the other hand, those tendencies that encourage morally reprehensible behaviours, such as aggressiveness, can be restrained. Furthermore, a genetic imprint of this educational activity can be achieved through proactive epigenesis. Finally, neuroscience highlights the close relationship between reasoning and emotions, therefore, it is necessary for education to take this fact into account, in order to foster moral agents' capacity to combine emotional responses to their environment with a critical analysis of the ethical codes that govern behaviour. Thereby, it becomes possible to narrow the distance between the codes that individuals find naturally inscribed in them and the universalist moral norms and values that ethics and reason seek to establish.

References

Apel, K. O. 1985. El a priori de la comunidad de comunicación y los fundamentos de la ética. En Apel, K. O., *La transformación de la filosofía* (Vol. II), 341–413. Madrid: Taurus.

Aranguren, J.L.L. 1994. *Ética*. Madrid: Alianza Editorial.

Ayala, F.J. 1987. The biological roots of morality. *Biology and Philosophy* 2 (3): 235–252.

Ayala, F.J., and C.J. Cela Conde. 2001. La filogénesis de la moral. In *Senderos de la evolución humana*, ed. F.J. Ayala and C.J. Cela Conde, 517–538. Madrid: Alianza Editorial.

Cayuela, A. 2012. The «evil ideology» description and critique of a policy against human nature (el «mal ideológico»: descripción y crítica de una política contraria a la naturaleza humana). *Cuadernos de Bioética XXIII* 2012/1a: 71–82.

Churchland, P. 2013. *El cerebro moral. Lo que la neurociencia nos dice sobre la moralidad*. Barcelona: Paidós.

Colmenares, F. 2005. De laberintos sociales y de cómo salir de ellos: inteligencia social. In *Existo, luego pienso: Los Primates y la Evolución de la Inteligencia Humana*, ed. F. Guillén-Salazar, 75–128. Madrid: Ateles.

Cortina, A. 2007. *Ética mínima: Introducción a la filosofía práctica*. Madrid: Tecnos.

———. 2009. *Ética de la razón cordial: educar en la ciudadanía en el siglo XXI*. Oviedo: Nobel, D. L.

———. 2012. *Neuroética y neuropolítica: Sugerencias para la educación moral*. Madrid: Tecnos.

———. 2017. *Aporofobia, el rechazo al pobre. Un desafío para la democracia*. Barcelona: Paidós.

Damasio, A. 2005. *En busca de Spinoza: Neurobiología de la emoción y los sentimientos*. Barcelona: Crítica.

———. 2008. *El error de Descartes: La emoción, la razón y el cerebro humano*. Barcelona: Crítica.

Darwin, C. 1979. *El origen del hombre y la selección en relación al sexo*. Madrid: EDAF.

Einsengberg, N. 2000. Emotion, regulation and moral development. *Annual Review of Psichology* 51: 665–697.

Evers, K. 2010. *Neuroética. Cuando la materia se despierta*. Madrid: Katz Editores.

———. 2015. Can we be epigenetically proactive? In *Open MIND: 13(T)*, ed. T. Metzinger and J.M. Windt. Frankfurt am Main: MIND Group.

Forbes, C.E., and J. Grafman. 2010. The role of the human prefrontal cortex in social cognition and moral judgment. *Annual Review of Neuroscience* 33: 299–324.

García-Marzá, D. 1992. *Ética de la justicia. J. Habermas y la ética del discurso*. Madrid: Tecnos.

Goldberg, E. 2002. *El cerebro ejecutivo: lóbulos frontales y mente civilizada*. Barcelona: Crítica.

González-Esteban, E. 2013. Formación ética de los profesionales. Forjando el interés desde la razón y la emoción. *RIO. Revista Internacional de Organizaciones* 10: 21–40.

Greene, J.D. 2003. From neural is to moral ought: What are the moral implications of neuroscientific moral psychology? *Nature Reviews of Neuroscience* 4: 847–850.

———. 2013. *Moral tribes. Emotion, reason and the gap between us and them*. New York: Pinguin.

Habermas, J. 2006. *Entre naturalismo y religión*. Barcelona: Paidós.

———. 2010. *Teoría de la acción comunicativa*. Madrid: Trotta.

Haidt, J. 2001. The emotional dog and its rational tail: A social intuitionist approach to moral judgment. *Psychological Review* 108 (4): 814–834.

———. 2013. *The righteous mind: Why good people are divided by politics and religion*. London: Penguin Books.

Hauser, M.D. 2008. *La mente moral. Cómo la naturaleza ha desarrollado nuestro sentido del bien y del mal*. Barcelona: Paidós.

Honneth, A. 1997. *La lucha por el reconocimiento: por una gramática moral de los conflictos sociales*. Barcelona: Crítica.

Jácomo, A., and D. Serrão. 2013. Implications of neuroplasticity in neuroethics. In *Jerusalém proceedings of the Jerusalem international conference on Neural plasticity and cognitive modifiability*, 73–78. Italia: Medimond.

Korsgaard, C. 2000. *Las fuentes de la normatividad*. México: Universidad Nacional Autónoma de México, Instituto de Investigaciones Filosóficas.
Le Doux, J.E. 1999. *El cerebro emocional*. Barcelona: Ariel.
Libet, B. 1999. Do we have free will? *Journal of Consciousness Studies* 6 (8–9): 47–57(11).
Moll, J., R. de Oliveira-Souza, I.E. Bramati, and J. Grafman. 2002a. Functional networks in emotional moral and nonmoral social judgments. *NeuroImage* 16: 696–703.
Moll, J., R. de Oliveira-Souza, P.J. Eslinger, I.E. Bramati, J. Mourao-Miranda, P.A. Andreiuolo, and L. Pessoa. 2002b. The neural correlates of moral sensitivity: A functional magnetic resonance imaging investigation of basic and moral emotions. *The Journal of Neuroscience* 22: 2730–2736.
Moore, G.E. 1983. *Principia ethica*. México: Universidad Nacional Autónoma de México.
Morgado, I. 2010. *Emociones e inteligencia social*. Barcelona: Ariel.
Nichols, S. 2005. Innateness and moral psychology. In *The innate mind: Structure and contents*, ed. P. Carruters, S. Laurence, and S. Stich, 353–430. New York: Oxford University Press.
Ortiz de Landázuri, C. 2016. El error neurocientífico de Descartes, entre Spinoza y Tomás de Aquino. El debate entre Damasio y Stump sobre el materialism eliminativo en la neurociencia, neuropolítica y neuroeconomía. *Recercare* 18: 107–133.
Pallarés-Domínguez, D A. Richart. 2018. Neuroética y neuroeducación: repensando la relación entre las neurociencias y las ciencias sociales. *Recerca* 22.
Pallarés-Domínguez, D. 2013. Críticas y orientaciones para el estudio en neuroética. *Recercare* 13: 85–102.
———. 2016. My brain made me not do it: An emergentist interprestation of Benjamin Libet. *Ramon Llull Journal of Applied Ethics* 7: 121–141.
Prinz, J. 2011. Where do morals come from? In *Empirical informed ethics: Morality between facts and norms*, ed. M. Christen et al. Springer.
Reverter Bañón, S. 2016. Reflexión crítica frente al neurosexismo. *Pensamiento* 72 (273): 959–979.
Reverter Bañón, S., and M. Medina-Vicent. 2018. Neuroeducación para la democracia y la igualdad. In *Juicio moral y democracia. retos de la ética y la filosofía política*, ed. A. Richart, M. García- Granero, C. Ortega Esquembre, and L. de Tienda Palop, 314–325. Granada: Comares.
Richart, A. 2016. El origen evolutivo de la agencia moral y sus implicaciones para la ética. *Pensamiento* 72 (273): 849–864.
Rizzolatti, G., and C. Sinigaglia. 2006. *Las neuronas espejo. Los mecanismos de la empatía emocional*. Barcelona: Paidós.
Roskies, A.L. 2010a. How does neuroscience affect our conception of volition? *Annual Review of Neuroscience* 33: 109–130.
———. 2010b. Why Libet's studies don't pose a threat to free will. In *Conscious will and responsibility*, ed. W. En Sinnott-Armstrong and L. Nadel, 11–22. Ney York: Oxford University Press.
Salles, A. 2015. Rationality and the moral significance of emotions. In *Inherent and instrumental values: Excursions in value inquiry*, 89–99. Lanham: University Press of America.
———. 2017a. Social neuroscience and neuroethics: A fruitful synergy. In *Neuroscience and social science: The missing link*, ed. A. Ibáñez et al., 531–546. Cham: Springer.
———. 2017b. Proactive epigenesis and ethics. *EMBO Reports* 18 (8): 1271.
Teehan, J., and C. diCaro. 2004. On the naturalistic fallacy: A conceptual basis for evolutionary ethics. *Evolutionary Psychology* 2: 32–46.
Tomasello, M., and A. Vaish. 2013. Origins of human morality. *Annual Review of Psychology* 64: 231–255.
Waal, F. 1997. *Bien natural: Los orígenes del bien y del mal en humanos y otros animales*. Barcelona: Herder.
———. 2007. *Primates y filósofos: La evolución de la moral del simio al hombre*. Barcelona: Paidós.

Chapter 4
Moral Neuroeducation, Ethics of Justice and Pluralism

César Ortega-Esquembre

4.1 Introduction

In order to design a model of moral neuroeducation that is compatible with the principles and values of a pluralistic society a number of challenges and risks have to be addressed at various conceptual levels. From the outset three core themes must be highlighted: firstly, the neurophysiological mechanisms underlying the processes by which moral skills are acquired; secondly, the strategies employed to exploit this biological knowledge for the purpose of designing moral education models; and finally, the specific ethical model that underlies the tasks undertaken by moral neuroeducation. The first theme is concerned with what may be broadly termed neuroethics, the second with moral neuroeducation, and the third belongs to the field of moral and political philosophy.

While, it is clear that these three areas of study are of critical importance to ensure that any model of moral neuroeducation created is done so thoroughly, the fundamental issues is that any proposed type of moral neuroeducation will be defined by the ethical foundation it is based on. Thereby, the moral-philosophical task has theoretical precedence over the neuroeducational task. If an educator draws on neuroscientific advances for the design of strategies to be employed in the training of moral competences, it is evident that they should be equipped with a core set of competences, norms or moral principles upon which to base their teaching. A knowledge of how humanity establishes norms and moral principles, and whether they rely upon sufficiently well-founded criteria, goes beyond the remit of both education theory and neuroscience, hence the need for this chapter's focus on moral

The author's research has been supported by a FPU contract-fellowship from Spain's Ministerio de Educación, Cultura y Deporte (Reference: FPU13/00631).

C. Ortega-Esquembre (✉)
Universidad de Valencia, Valencia, Spain
e-mail: Cesar.Ortega@uv.es

P. Calvo, J. Gracia-Calandín (eds.), *Moral Neuroeducation for a Democratic and Pluralistic Society*, https://doi.org/10.1007/978-3-030-22562-9_4

and political philosophy. Drawing on the conceptual framework of Adela Cortina's ethics of cordial reason, the discussion that follows, will underscore the crucial importance of cognitivist, deontological and "liberal" factors for the development of moral neuroeducation.

There is the possibility that a moral neuroeducation project may awaken suspicions of indoctrination and the only way this can be countered is by emphasising how this form of education has nothing to do with any particularistic content, namely with the "ethical maximum" (Cortina 1986), "comprehensive doctrines" (Rawls 1990) or "felicitous forms of life" (Habermas 1983). Instead, it is exclusively concerned with those norms and values that are universalizable, as well as the principles and justification procedures that underpin such a universalisation. In other words, neuroeducation is to be placed within the framework of an "ethics of justice" (García-Marzá 1992), as it is committed to the specific ethical challenge of modernity: the defence of a universalist sense of ethics that wholly respects axiological pluralism.

In order to elaborate on this, the meaning of the term "moral neuroeducation", and the novel interdisciplinary approach to education it entails, must first be clarified. Likewise, the aforementioned risk of indoctrination requires further consideration in conjunction with the necessary precautions to be taken in politico-moral and philosophical terms. Attention will then be turned to a survey of the main theoretical sources –John Rawls, Jürgen Habermas, Karl Otto Apel and Adela Cortina–, all based on a shared Kantian foundation, in order to identify the most suitable ethical model for moral neuroeducation. To conclude it will be demonstrated how this model successfully dispels any concerns over indoctrination.

4.2 Moral Neuroeducation and Its Risks

The impressive developments achieved by the natural sciences have led to varying degrees of engagement with other disciplines, such as moral philosophy, which are traditionally considering as constituting what is known as the *Geisteswissenschaften*. So long as this scholarly development does not give rise to the *colonisation* of moral philosophy and its sister disciplines by the methodology of the natural sciences, and instead takes the form of an interdisciplinary collaboration intended to clarify issues that resist any one-dimensional explanation, it may be argued that this phenomenon promises to be exceptionally fruitful for the advance of scientific knowledge. In this regard moral neuroeducation undoubtedly represents one of the most interesting outcomes of this recent intellectual trend, as its emergence combines elements from three previously disconnected fields of knowledge: cognitive neuroscience, moral philosophy and education.

Given the somewhat surprising lack of scholarly literature on moral neuroeducation, the best way of providing an operational definition of the discipline is to briefly explain what is meant by "neuroeducation", and then to apply this concept to a more specific moral educational task. Neuroeducation, also known as "educational neuroscience" (Bowers 2016), is a transdiscipline that seeks to draw on studies of the human brain in order to design new educational practices. The main "Neuroscience and education argument", addressed by John T. Bruer in his seminal article "Education and the Brain: A Bridge Too Far" (Bruer 1997), claims that the contributions of neuroscience can shed light on the brain activity underlying learning process, and that this "is relevant to education and can lead to improve teaching and learning" (Howard-Jones 2016: 620).

Naturally, this argument takes as its point of departure the scientifically validated thesis of neuronal plasticity: brain structure and neuronal connectivity can be deliberately modified through education (Koizumi 2005; Tokuhama-Espinosa 2008; Marina 2012). As María José Codina has observed, "the aim of neuroeducation, unlike the aims of cognitive neuroscience and neuropsychology, is not only to understand how human beings learn better, but also to consider the way in which they can be taught to maximise their potential" (Codina 2014: 27). Neuroeducation is not restricted to a descriptive level, it also addresses both a prescriptive level (the application of neuroscientific learning for the design of specific pedagogical methodologies) and a therapeutic level (in which it is used to tackle learning difficulties) (Pallarés-Domínguez 2016).

Despite its very recent development, this new transdiscipline has given rise to a great deal of controversy, and as yet no consensus has been reached on the veracity of the "neuroscience and education argument". Some authors argue that, it is probable that in the future neuroscience will play an important role in the design of educational practices; so far it has not contributed to the development of any new useful teaching practices (Bowers 2016: 600; Bruer 1997). According to this view, whose proponents seek to debunk the so-called neuromyths about process learning (Geake 2008), it is only through the mediation of other disciplines, such as cognitive psychology, that the view of neurobiology as a "bridge too far" for education may be addressed (Bruer 1997, 2016). For the present, I wish to set aside this controversy, which other chapters in this book analyse in greater depth, and instead I want to turn to the application of this provisional definition of neuroeducation to the more specific field of moral neuroeducation.

Although neuroeducation, as defined above, applies above all to the design of learning programmes related to reading or the acquisition of mathematical competences (Ansari et al. 2012), the truth is, its achievements can also be applied to the field of moral competences (Gracia-Calandín 2018). If neuroeducation seeks to draw on studies of the brain activity underlying learning processes for the design of new educational practices, then moral neuroeducation must be based on the neuroscientific study of the brain mechanisms involved in learning about moral norms, moral actions and moral judgements, which could in turn be used to design educational programmes related to moral behaviour. In line with this, moral neuroeducation can only emerge as the result of the conjunction between moral education and neuroethics (Cortina 2011); the latter includes research undertaken in evolutionary anthropology studies on the emergence of moral systems (Alexander 1987), as well as on moral agency (Joyce 2014; Richart 2017) and moral consciousness (Cortina 2013).

Nonetheless, moral educational tasks are always threatened by the risk of indoctrination, as can be observed in the current political debate on the design of civic and ethical education curricula. Although moral education must foster an appropriate ethical development for children and teenagers, the truth is that any one educator cannot determine alone, and according to his/her own criteria, what is "ethically appropriate". As Adela Cortina observes, "the educator has no right to inculcate his/her own ideal of happiness as if it were an universalizable ideal" (Cortina 1993: 215), because in liberal democracies based on the principle of pluralism, only the ethical minimum that is universally recognised, in accordance with principle of rule according to higher law, should be made mandatory for all citizens, as opposed to the particularistic ideal of a felicitous life. If we do not think carefully about the contents that educators can legitimately transmit to students, there is a risk of moral education degenerating into indoctrination. Obviously, neuroscientific research cannot solve this problem itself, and indeed may make the situation worse: if this problem is not solved at the level of moral and political philosophy, then neuroeducation would merely reinforce a more efficient way of indoctrination because, through this transdiscipline we would come to depend on neuroscientific advances to both rationalise education processes and maximise their results. Or to put it another way: ethics for moral neuroeducation must not be limited to a naturalistic paradigm (Conill 2019; Gracia-Calandín 2019). In terms of moral philosophy, it is deontological ethics that offers the most convincing solution to this problem, while it is political liberalism that provides a means to address this issue in the context of political philosophy. Therefore, over the next two sections, an analysis is undertaken of the most relevant elements of these two branches of thought, and it is shown how they support the case for moral neuroeducation, above all through the cognitive, deontological and liberal approaches they provide.

4.3 Cognitivism, Deontology and Political Liberalism

From a non-cognitivist ethical perspective, that is a view based on the relativistic or sceptical thesis according to which it is not possible to speak about normative rightness in the same sense as we speak about truth, the problem of a potential indoctrination risk is, admittedly, difficult to overcome. If we are not prepared to admit the existence of norms and moral principles that not only have validity in a certain "justification context" (Rorty 1989, 1991), but also a universal validity within an "ideal community of communication" (Apel 1976a, b), then it is not clear what moral educational content would be universally acceptable for the design of educational programmes. If we wish to address the debate on indoctrination risks with any prospect of success, then there should be at least a normative point of view from which to denounce the risks of particularism.

Approaches based on ethical cognitivism have tackled this problem by opposing the aforementioned relativistic thesis, above all through the various models of moral and political philosophy developed by Karl Otto Apel, Adela Cortina, John Rawls,

Jürgen Habermas and Lawrence Kohlberg, who have all drawn on the work of Kant. The core idea underpinning ethical cognitivism, regardless of the differences between these various models, is that moral norms are susceptible to validity or rationality in a similar sense as scientific statements are.

Discourse ethics, perhaps the most developed programme of cognitivist ethics, seeks to support this thesis through the development of a procedure to justify moral and legal norms (normative rightness). This procedure is analogous to the procedure for the justification of scientific statements (truth), and both are based on the idea of a consensus, or a discourse that is guided by rational consensus and endeavours to embrace the entire range of validity claims (*Geltungsansprüche*). In other words, these two procedures are based on a rational discourse through which a resolution is offered to full range of potential criticism and counter-criticism that participants may raise when seeking to justify a norm (Habermas 1983; Apel 1976a, b). As opposed to radical contextualism in epistemology (Richard Rorty), communitarianism in moral philosophy (MacIntyre) and decisionism in legal philosophy (Niklas Luhmann or Carl Schmitt), some authors like Jürgen Habermas (1983), Karl Otto Apel (1976a, b), Albrecht Wellmer (1986), Adela Cortina (1990) and Domingo García-Marzá (1992), have attempted to prove that the idea of "normative rightness" (*Richtigkeit*) contains an implied reference to the idea of "truth" (*Wahrheit*). The normative consensus reached argumentatively, under "violence-free" discourse conditions, is the expression of a rational will, whose rationality depends on the formal properties of the discourse itself. In the same way that false scientific statements can be distinguished from those that are true, correct moral norms that protect the general interest can be distinguished from flawed norms, whose existence is only factually based (Habermas 1998). The definition of a fact as "true", as well as the definition of a norm as "correct", mean that any such a fact or norm "is argumentatively justifiable for all those equipped with the communicative competence" (Cortina 1990: 112). In order to define the meaning of this "communicative competence", that is, this ability to raise validity claims and to defend them against "interaction partners", Jürgen Habermas and Karl Otto Apel designed a very sophisticated theory called "formal pragmatics" (Habermas 1976; Apel 1976a, b). Regrettably, it is beyond the scope of this chapter to address this theory.

Habermas distinguishes between theoretical and practical discourse. Theoretical discourse, based on the formation of hypotheses, endeavours to eliminate those scientific beliefs that fail to pass the rigorous test of refutation within the "community of inquiry" (Peirce). Practical discourse likewise operates in a critical manner, but with regard to the naive mode of existence adopted by norms, and this is termed "social" or "factual" validity. Just as is the case with Habermas's Consensus Theory of Truth (Habermas 1973), the claim for a statement's validity, and the possibility of any such valid statement claiming "discursive redemption" (*diskursive Einlösung*), suggests that "truth" has an unconditional nature. However, in discourse ethics, the idea of the discursive redemption of a normative validity claim signals an idea of "rightness", a universalist status; this is precisely the status Kant that claimed conformed with his notion of a "moral point of view". More recently, Kohlberg attempted to address this issue with regard to his idea of a "post-conventional" level of moral development. Naturally unlike "truth"-related validity claims, validity claims related to normative rightness do not refer to an objective world, but instead to a social world that is the result of our very interactions. As a result, normative rightness is unable to appeal to an objective space outside of discourse; that is, it is unable to go beyond the intriguing plot of reality called "language".

In contrast to Habermas's discourse ethics, John Rawls's discussion of the ethical cognitivism within political liberalism locates its foundations in the theoretical construct he terms the "original position", and which he refers to as a hypothetical situation that satisfies the formal conditions to produce valid agreements (Rawls 1990). From his early works in the 1950s, Rawls has attempted to challenge non-cognitivist views of ethics. Akin to Habermas and Apel's discourse ethics, and also to Cortina's ethics of cordial reason, Rawls provides a rational justification of the moral principles of justice. According to Rawls, once religious-metaphysical world-views have lost their cohesive force, the rightness, or what Rawls also terms, the "objectivity" of moral norms will no longer correspond with any objective moral order. Instead, principles of political justice will be established through political constructivism, which is the category used by Rawls; these principles being the result of a constructive procedure between rational and reasonable agents. The objectivity of a moral norm or judgement does not require a correlation between these norms or judgements and the world—a correlation, which would in any case be encompassed by theoretical reason—, instead the sole requirement is simply that sufficient reasons are offered and considered by citizens. Normative rightness is, thus, based on the rational acceptability of all concerned, an acceptability that Rawls represents by his impressive operationalisation of classical contractualism through the formulation of the figure of an "original position". The formal properties of this original position are characterised by the fact that none of the participants in the deliberation process have any knowledge regarding their social status, particular abilities or position within a social order, what is termed the "veil of ignorance"; this guarantees that any agreements reached have a rational character, that is, such agreements represent universal interests (Rawls 1990).

The notions of ethical cognitivism set out in the idea of reaching an agreement in both the "ideal speech situation" and the "original position", have to be developed in order to offer insight into what kind of norms and moral principles are universalizable. For proposals of both discourse ethics and political liberalism to make sense, a clear separation must be made between the ethical minimum and the particularistic ideals of a felicitous life. Therefore, the model that moral neuroeducation seeks, in order to overcome any risk of indoctrination, must be cognitivist, as well as deontological with regard to its moral-philosophical dimension, and liberal in terms of its political-philosophical dimension.

As opposed to other approaches, such as utilitarianism and communitarianism, discourse ethics endeavours to develop a procedural principle of universalisation. This procedural principle is designed to resolve issues of justice (*the right*), not issues of happiness (*the good*), whereby the "priority of the right [is established] over the good". However, in contrast to the fate of the original Kantian model, the universalisation process discussed here does not take place in what Apel termed a "methodological solipsismus" (Apel 1996), or through a "monologicial" verification expressed as a categorical imperative. Instead, it is conducted through a dialogue between all those who are potentially affected by "the consequences and side effects" of the norm's general observance. As Thomas McCarthy observes in his interpretation of this fundamental development in Kantian ethics, the emphasis shifts "from what each can will without contradiction to be a general law, to what all can will in agreement to be a universal norm" (McCarthy 1978: 326).

In any event however, practical discourse does not play the role of generating valid norms, but the role of verifying or rejecting the validity of the strictly moral norms that already exist in the lifeworld (*Lebenswelt*); these norms have nothing to do with "the good", but with "the right", that is, they are susceptible to universalisation regardless of any particular form of life. From this moral point of view, practical discourse provides a means to select, from among a set of evaluative issues, only those problems that can be solved by reference to the general interest. Moral judgements aim to question the validity or invalidity of moral norms, not the success or failure of a form of life. Although these forms of life should not contradict moral standards, "their substance cannot be justified from universalistic points of view" (Habermas 1981: 168). Adela Cortina explains this basic problem of moral philosophy by distinguishing between a minimalist sense of ethics (justice) and a maximalist one (happiness) (Cortina 1986).

Naturally, such an approach to deontological moral philosophy has to cope with the problem of making space for "the good" within a theory about "the right". If it does not, then any such approach to moral philosophy would lose one of the building blocks of ethics (Cortina 1990). In a 1986 article on "stage 6" of Kohlberg's Moral Development Theory, Jürgen Habermas sought to solve this problem by articulating the principles of "justice" and "solidarity" (Habermas 1986). Habermas based his discussion on George Herbert Mead's Theory of Self, and specifically on Mead's thesis about individuation through socialisation. He demonstrated that the greater or lesser scale of any process of a subject's individuation is reflected in the greater or lesser extent to which the same subject enters into a network of mutual dependences and "need for protection". The reason for this is that an individual's self-configuration can only be achieved through that individual being recognised (*anerkennen*) by others. According to this dialectic, any one individual has to consider the perspective of a second person who becomes a foreigner (*entfremden*), so to speak. Thereby an individual becomes the "alter ego" of an "alter ego". It was within this context that Habermas defined morality as "the protective institution that compensates for a constitutional precariousness implicit in the sociocultural form of life itself" (Habermas 1991b). Yet, were that to be the case, then the integrity of an individual could only be guaranteed by preserving the integrity of a shared lifeworld. Habermas concluded that a further dimension of the equal treatment of individuals is not benevolence, but solidarity (Habermas 1986).

Many authors consider Habermas's attempt to correct his own deontological excesses insufficient. Adela Cortina, whose own ethical model employs the same deontological framework, identifies the idea of "compassion" as the "other of justice". In her view, what generates moral obligation, is not only *respect* for human dignity, but also *compassion* for the constitutional vulnerability of humanity.

> There are at least two moral voices: the voice of justice, which consists of judging what is right and wrong from a universal perspective, and the voice of compassion for those who need help. [...]. After all, there is no true justice without a sense of solidarity with the less fortunate, nor true solidarity unless it is founded on justice (Cortina 2007: 151).

In regard to the idea of the "other of justice", Axel Honneth has examined certain branches of recent French philosophy. Honneth has demonstrated how Lévinas and Derrida's contributions to the field of moral philosophy question the validity of the Kantian theoretical framework. Honneth identified in Derrida's model of moral philosophy two different classes of moral relationship: firstly, that of "friendship", in which another individual appears "in the role of the concrete, unrepresentable individual person" (Honneth 2000); and secondly, a relationship of equal treatment, in which the other individual appears "in the role of the generalised other [...] as a moral person just like everyone else". Honneth claims the relevance of the first kind of moral relationship, based on the idea of "affection", to be "the other of justice". His conceptual sense of "affection" is not simply a synonym of Habermas's concept of "solidarity", because, in the latter case, "compassion" for others does not refer asymmetrically to a certain individual, but universally to all human beings.

A further contribution to the critique of deontological aspects of discourse ethics, has been made recently in the areas of communitarian and feminist thought (MacIntyre 1981; Taylor 1991; Gilligan 2003; Benhabib 1992; Fraser 1999). MacIntyre, working within an Aristotelian framework, has argued that the price to pay for the Kantian moral point of view is a painful abandonment of the genuine content of ethics, namely guidance towards the good life and happiness within a particular community (MacIntyre 1981). Charles Taylor, working in the context of Romantic, Hegelian and humanist thought, singled out for critique what he termed as "the atomism of [the] moral and political philosophy" of Modernity, on in other words the atomistic individualism of liberal theories of contract. The core thesis of his argument is that moral norms of justice are not based on an impartial point of view, but on certain "horizons of significance" and "ideas of the good", which Taylor calls "strong evaluations" (Taylor 1991). Subsequently, Nancy Fraser, working from a feminist perspective, questioned the overly rigid application of the principle of neutrality, or in other words an excessively prescriptive separation between "public issues" (justice) and "private issues" (good life or happiness). She argues that this would exclude from public discourse precisely those phenomena that were, traditionally, considered to be private issues, but are instead sources of oppression (Fraser 1999).

Habermas and Apel pay heed to the communitarian critique, although they responded to MacIntyre and Taylor's critical evaluation of the "social contract" and "universal point of view" as "modern abstractions" by arguing that the denial of these achievements would come at the cost of political regression (Habermas 1991a). Continuing this line of argument, Habermas argued that communitarian criticism of liberal thought does not apply to discourse ethics as this model takes a middle ground somewhere between "liberal thinking"—the deontological view of morality, freedom and law inherited from Kant—and "communitarian thinking"—the intersubjective view of the individuation process as socialisation inherited from Hegel—(Habermas 1991). However, Habermas did take into account Fraser's criticism and endeavoured to overcome the problem by addressing the "public/private" distinction from two points of view: one the one hand, the informal shaping of public opinion, and on the other, political regulation (Habermas 1992).

Within this debate prompted by the communitarian critique, John Rawls took a clearly liberal position but, in turn, he corrected the individualistic deficits of liberalism with elements of a social democratic nature. The foundation of his well-known political philosophy is the "fact of reasonable pluralism", which attests to the existence of a diversity of opposing and irreconcilable reasonable doctrines; Rawls attempted to answer the key question of political liberalism: "how is it possible for there to exist over time a just and stable society of free and equal citizens, who remain profoundly divided by reasonable religious, philosophical, and moral doctrines?" (Rawls 2005: 4). In order to answer this question, Rawls designed a liberal political conception of justice, embodied in the two famous principles of justice as fairness: the principles of liberty and difference. Rawls's idea of a liberal political conception of justice involves three essential features, which are of particular relevance for the present discussion. Firstly, Rawls's idea of justice is applied to what he termed "the basic structure of society", or in other words, the main political, social, and economic institutions. Secondly, Rawls presents justice as operating from a "freestanding view" that is independent of all the comprehensive doctrines, despite the fact that his idea of justice "fits into, and can be supported by, various reasonable comprehensive doctrines" (Rawls 2005: 12) Finally, the content of Rawlsian justice is articulated "in terms of certain fundamental ideas seen as implicit in the public culture of a democratic society" (Rawls 2005: 13).

The source of disagreement between reasonable citizens, that is, between citizens willing to organise their social life in a just manner, arises in what Rawls called "burdens of judgement", which he refers to as follows: "the many hazards involved in the correct (and conscientious) exercise of our powers of reason and judgement in the ordinary course of political life" (Rawls 2005: 55). The citizens' capacity for reason means that, even when embracing different comprehensive doctrines, they are able to recognise that their own doctrine does not have to be imposed by the state on all the other members of society. Political liberalism, therefore, seeks to establish the terms for an equitable social cooperation, or in other words a political conception of justice, which all citizens, regardless of their comprehensive doctrines, recognise as legitimate. To explain how it is possible for this process to successfully and peacefully occur, Rawls made recourse to the idea of an "overlapping consensus"; that is, a consensus according to which each comprehensive doctrine accepts the agreed political concept of justice, albeit from its own perspective. Yet, for this to happen, the political concept of justice needs to be restricted to a small number of issues capable of being generally accepted. In one of the most meaningful passages of contemporary political philosophy, John Rawls summarises this deontological core of political liberalism as follows:

> We do best not to assume that there exist generally acceptable answers for all or even many questions of political justice. Rather, we must be prepared to accept the fact that only a few questions we are moved to ask can be satisfactorily resolved. Political wisdom consists in identifying those few, and among them the most urgent. [...] A political conception is at best but a guiding framework of deliberation and reflection which helps us reach political agreement on at least the constitutional essentials and the basic questions of justice (Rawls 2005: 156).

Rawls thus defends what he termed the "priority of the right over the good". Justice as fairness, understood as a political concept of justice, seeks a public basis of justification that is sufficient to form a focus for an entwined consensus of comprehensive doctrines. Such a public basis of justification is encountered in the aforementioned "constitutional essentials" and the "basic questions of justice"; that is, exactly those issues which the state can legitimately impose by using coercive power.

4.4 Ethical Minimalism: An Antidote to Indoctrination

The succinct survey of the cognitivist, deontological and liberal elements of the Kantian legacy of moral and political philosophy also offers a convincing solution to the problem raised by moral neuroeducation, as outlined at the outset of this chapter. Drawing on this survey I would argue that there is a *kind* of ethical content that certain models of moral neuroeducation may take up without risking any violation of the principle of axiological pluralism. It is only with the clarification provided by moral and political philosophy that the progress made by the natural sciences can be correctly integrated into education. However, this issue in turn raises others, including the oft-recurring case of science's misunderstanding of itself, which tends to be both "inflationary", with regard to its own competences, and "deflationary", in relation to the various "concepts of experience" (Conill 1998). Such a misunderstanding triggers the alarm raised by Habermas's ideology critique (Habermas 1968a, b), and it is only insofar as science is recognised as a historically determined form of knowledge that it can eliminate any suspicion of it tending towards ideology (Horkheimer 1937; Habermas 1968a, b).

In addition to an ideological critique of science, masterfully addressed through philosophical hermeneutics and the Critical Theory of Society, a further idea may be upheld, one supported by the thesis of moral cognitivism: as is the case for all the other areas of cultural knowledge—science, law and art—, the contents of ethics can be perfectly transmitted through educational programmes. Just as mathematics teachers teach *mathematical truth* and history teachers teach *historical truth*, teachers of ethics can teach what is *normatively correct*. If scientific truth is based on what, at present, the research community recognises as "true", the correction of moral norms is based on what, under the conditions of a non-coercive and inclusive argumentative situation, all those potentially affected by those norms—that is, all human beings—, could recognise as being "correct". Of course, such a universalistic agreement can only be reached on very specific issues: only on those issues that deontological ethics identifies as susceptible to universalisation, and political liberalism identifies as being legitimately manageable for the state, while not risking any threat to the principle of axiological neutrality. However, no generation starts from scratch, so to speak, when it comes to learning moral norms. Rather the opposite is the case; there are historic achievements that systematically support the thesis of moral progress, the accomplishments which must be transmitted through educational programmes.

With such a model of ethics, the aims of moral neuroeducation (i.e., the attempt to employ neuroscientific advances in the design of strategies relating to the training of moral competences) would be freed of any risk of indoctrination. This is a risk that cognitive science cannot solve for itself, and it is one that can increase to dangerous levels if the appropriate remedial plans are not applied at the level of moral and political philosophy. The distinction between the right and the good, that is, between the minimum ethical foundation that serves as the normative basis of both the rule according to higher law, and the particularistic ideal of felicitous life upon which such a state can only intervene in a negative manner, avoids any risk of an indoctrination, as this would be nothing other than an attempt to present concerns for the good life *as if they were* issues concerns for justice.

In regard to this Adela Cortina has stated that, "education must aim to empower children to distinguish between community or conventional rules and universalistic principles, as this allows us to criticise even community norms" (Cortina 1993: 217). The above statement is especially significant for the discussion at hand, as it places emphasis on the agency of the educated subject. Properly founded moral norms, such as those that can gain universal consent, above all human rights, can and should be transmitted through formal and informal education. However, a compelling issue arises with regard to the process of transmission, namely that once the procedural principle of transmission is integrated into a "dialogical ethos" (Cortina 1993: 221) it enables those who are affected by moral norms to speak about the content of the norms that affect them. Thereby, only through an education guided by the principles of autonomy and critique, can the risks analysed here be effectively avoided.

4.5 Conclusions

The quest for interdisciplinarity, which has been a constant feature of the history of scientific knowledge, is not only an intrinsic hazard to science, but is also the most relevant and effective manner to advance scientific knowledge. However, any such progress is perverted when aspects of scientific knowledge contributed to the new interdiscipline, or transdiscipline, deceptively claims epistemic privileges; the result is that the aims underpinning the diverse contributions of knowledge to the new interdisciplinary model are threatened.

Moral neuroeducation is a good example of a transdiscipline that, when understood correctly, can solve problems that resist being resolved from any one perspective. A model of ethics that renounces the advances made by the different scientific disciplines, amidst which the neurosciences undoubtedly play a very important role, would not genuinely belong to its contemporary era. However, any such ethical model must also have a clear knowledge of its role within the new "division of labour" and must vigorously defend that role from any risk of being colonised by the other fields of knowledge. Within the framework of moral neuroeducation, moral and political philosophy should streamline the *kind* of ethics that is appropriate for the aims of neuroscience by maximising the results of moral education. If we are to remain faithful to the spirit of a rule according to higher law and liberal democracy, that is, maintaining a respect for the principle that any doctrine must assume the minimum standards of justice, then moral and political philosophy must fulfil the simple, but crucial task of explicitly formulating a set of appropriate procedures that can in turn be shaped by the moral norms which decide on their validity.

References

Alexander, Richard. 1987. *The biology of moral systems*. Hawthorne: Aldine de Gruyter.

Ansari, Daniel, Bert De Smedt, and Roland H. Grabner. 2012. Neuroeducation. A critical overview of an emerging field. *Neuroethics* V: 105–107.

Apel, Karl Otto. 1976a. Das Problem der philosophischen Letztbegründung im Lichte einer transzendentalen Sprachpragmatik: Versuch einer Metakritik des Kritischen Rationalismus. In *Sprache und Erkenntnis*, ed. B. Kanitscheider, 55–82. Innsbruck: Institut für Sprachwissenschaft der Universität.

———. 1976b. Sprechakttheorie und transzendentale Sprachpragmatik zur Frage ethischer Normen. In *Sprachpragmatik und Philosophie*, 10–103. Frankfurt a.M.: Suhrkamp.

———. 1996. Diskursethik als Verantwortungsethik –eine postmetaphysische Transformation der Ethik Kants. In *Kant in der Diskussion der Moderne*, ed. Y. Kato and G. Schönrich, 326–359. Frankfurt a.M: Suhrkamp.

Benhabib, Seyla. 1992. *Situating the self: Gender, community and postmodernism in contemporary ethics*. New York: Routledge.

Bowers, Jeffrey S. 2016. The practical and principled problems with educational neuroscience. *Psychological Review* CXXIII/5: 600–612.

Bruer, John T. 1997. Education and the brain: A bridge too far. *Educational Researcher* XXVI/8: 4–16.

———. 2016. Neuroeducación: un panorama desde el puente. *Propuesta Educativa* XLVI/25: 14–25.

Codina, María José. 2014. *Neuroeducación en virtudes cordiales*. Valencia: Universidad de Valencia.

Conill, Jesús. 1998. Concepciones de la experiencia. *Diálogo filosófico* XLI: 148–170.

———. 2019. Must ethics for moral neuroeducation be naturalistic? In *Moral neuroeducation for a democratic, and pluralistic society*, ed. Patrici Calvo and Javier Gracia-Calandín. Cham: Springer. [forthcoming].

Cortina, Adela. 1986. *Ética mínima*. Madrid: Tecnos.

———. 1990. *Ética sin moral*. Madrid: Tecnos.

———. 1993. *Ética aplicada y democracia radical*. Madrid: Tecnos.
———. 2007. *Ética de la razón cordial*. Oviedo: Nobel.
———. 2011. *Neuroética y neuropolítica*. Madrid: Tecnos.
———. 2013. La conciencia moral: Entre la naturaleza y la autonomía. *Cuadernos Salmantinos de filosofía* XL: 249–262.
Fraser, Nancy. 1999. Rethinking the public sphere: A contribution to the critique of actually existing democracy. In *Habermas and the public sphere*, ed. C. Calhoun, 109–142. Cambridge, MA: The MIT Press.
García-Marzá, Domingo. 1992. *Ética de la justicia*. Madrid: Tecnos.
Geake, John. 2008. Neuromythologies in education. *Educational Research* L/2: 123–133.
Gilligan, Carol. 2003. *In a different voice: psychological theory and women's development*. Cambridge: Harvard University Press.
Gracia-Calandín, Javier. 2018. El fin ético no naturalista de la neuroeducación. *Recerca. Revista de Pensament i Analisi* XXII: 51–68.
———. 2019. Non-naturalistic neuroethics for moral neuroeducation. In *Moral neuroeducation for a democratic, and pluralistic society*, ed. Patrici Calvo and Javier Gracia-Calandín. Cham: Springer. [forthcoming].
Habermas, Jürgen. 1968a. *Erkenntnis und Interesse*. Frankfurt a.M.: Suhrkamp.
———. 1968b. *Technik und Wissenschaft als «Ideologie»*. Frankfurt a.M.: Suhrkamp.
———. 1973. Auszug aus «Wahrheitstheorien». In *Wirklichkeit und Reflexion*, ed. H. Fahrenbach, 211–265. Pfüllingen: Neske.
———. 1976. Was heißt Universalpragmatik? In *Sprachpragmatik und Philosophie*, ed. K.O. Apel, 174–272. Frankfurt a.M.: Suhrkamp.
———. 1981. *Theorie des kommunikativen Handelns*. Vol. II. Frankfurt a.M.: Suhrkamp.
———. 1983. Diskursethik – Notizen zu einem Begründungsprogramm. In *Moralbewußtsein und kommunikatives Handeln*, 53–125. Frankfurt a.M.: Suhrkamp.
———. 1986. Gerechtigkeit und Solidarität. Zur Diskussion über "Stufe 6". In *Zur Bestimmung der Moral*, ed. W. Edelstein and G. Nummer Winkler, 291–318. Frankfurt a.M: Suhrkamp.
———. 1991a. Erläuterungen zur Diskursethik. In *Erläuterungen zur Diskursethik*, 119–226. Frankfurt a.M.: Suhrkamp.
———. 1991b. Treffen Hegels Einwände gegen Kant auch auf die Diskursethik zu? In *Erläuterungen zur Diskursethik*, 9–30. Frankfurt a.M: Suhrkamp.
———. 1992. *Faktizität und Geltung*. Frankfurt a.M.: Suhrkamp.
———. 1998. Richtigkeit versus Wahrheit. Zum Sinn der Sollgeltung moralischer Urteile und Normen. *Deutsche Zeitschrift für Philosophie* XLVI/2: 179–208.
Honneth, Axel. 2000. Das Andere der Gerechtigkeit. Habermas und die Herausforderung der poststrukturahstischen Ethik. In *Das Andere der Gerechtigkeit*, 133–170. Frankfurt a.M.: Suhrkamp.
Horkheimer, Max. 1937. Traditionelle und kritische Theorie. *Zeitschrift für Sozialforschung* VI: 245–294.
Howard-Jones, Paul. 2016. The principles and practices of educational neuroscience: Comment on Bowers. *Psychological Review* CXXIII/5: 620–627.
Joyce, Richard. 2014. The origins of moral judgment. In *Evolved morality. The biology and philosophy of human conscience*, ed. F.B. De Waal, P.S. Churchland, T. Pievani, and S. Parmigiani. Boston: BRIL.
Koizumi, Hideaki. 2005. Brain-science and education. In *Brain, science and education*. Saitama: Japan Science and Technology Agency.
MacIntyre, Alasdair. 1981. *After virtue: A study in moral theory*. London: Duckworth.
Marina, José Antonio. 2012. El diálogo entre Neurociencia y Educación. *Participación Educativa* 2012: 7–15.
McCarthy, Thomas. 1978. *The critical theory of Jürgen Habermas*. Cambridge: MIT Press.
Pallarés-Domínguez, Daniel. 2016. Neuroeducación en diálogo: neuromitos en el proceso de enseñanza-aprendizaje y en la educación moral. *Pensamiento* LXXII/273: 941–958.

Rawls, John. 1990. *A theory of justice*. Oxford: Oxford University Press.
———. 2005. *Political liberalism*. New York: Columbia University Press.
Richart, Andrés. 2017. El origen evolutivo de la agencia moral y sus implicaciones para la ética. *Pensamiento* LXXII/273: 849–864.
Rorty, Richard. 1989. *Contingency, irony, and solidarity*. Cambridge: Cambridge University Press.
———. 1991. *Objectivity, relativism, and truth*. Cambridge: Cambridge University Press.
Taylor, Charles. 1991. *The ethics of authenticity*. Cambridge: Harvard University Press.
Tokuhama-Espinosa, Tracey. 2008. *The scientifically substantiated art of teaching*. Minneapolis: Capella University.
Wellmer, Albrecht. 1986. *Ethik und Dialog*. Frankfurt a.M.: Suhrkamp.

Part II
Moral Emotions in Neuroeducation

Chapter 5
The Role of the Emotions in Moral Neuroeducation

Lydia de Tienda Palop

5.1 Introduction

A key trend in current moral education studies is the discussion of the links between moral education and emotional management. The emergence of this recent line of enquiry highlights the need to consider various areas of theoretical discussion in order to ensure that the teaching of moral education continues to advance towards its goal of achieving moral progress in society. First and foremost, it must be asked whether moral education and emotional management mean the same thing, and if not, what differences there are between them, however nuanced they may be. The relevance of this question is underscored by neuroscientific research, which frequently identifies the emergence of various emotions with moral behaviour. One factor underpinning this scientific view is the fact that emotions can easily be observed empirically, and current neuroscientific technology is sufficiently sophisticated to perform measurements and experiments with empirical samples. However, before proceeding a survey of this area of neuroeducational research is required in order to understand the specific nature of moral education. According to a range of key authors (Ansari et al. 2011, 2012; Campbell 2011; Marina 2012) neuroeducation is concerned with how studies of the brain can provide a key contribution to our understanding of education, and it seeks to develop a field of study concerned with generating new learning strategies and methodologies for education (Pallarés 2016). Through this new field of study links are forged between pedagogy and cognitive psychology, and likewise with naturalistic studies on the organic functioning of neural systems. The studies carried out to date in this field have already had an impact on the understanding of learning processes, information processing and also the development of cognitive abilities. However, moral education studies have certain characteristics that prevents it being assimilated into the study of cognitive

L. de Tienda Palop (✉)
Complutense University, Madrid, Spain
e-mail: lydiadet@ucm.es

P. Calvo, J. Gracia-Calandín (eds.), *Moral Neuroeducation for a Democratic and Pluralistic Society*, https://doi.org/10.1007/978-3-030-22562-9_5

learning processes and vice versa. While it is true that a range of authors have contributed to results from the field of neuroscience being applied to ethics, and those of ethics to neuroscience, it is essential to underscore the specific nature of moral education, as well as its connections with emotional management, and to do so a number of fundamental preliminary points need to be taken into account.

As Gracia-Calandín has made clear, there is considerable potential to combine neuroeducation and ethics, as the former is a field of study that encompasses multiple perspectives. Therefore, it is not the same to talk about the ethical implications of neuroeducation as it is to discuss those of ethical neuroeducation, or even of educational neuroethics, which are just some of the examples discussed by Gracia-Calandín (2018: 55). Nonetheless, they all develop a combined scope on neuroscience-education-ethics. In this regard Gracia-Calandín, drawing on the work of Roskies (2002: 21), has identified a fundamental distinction between the "ethics of neuroscience"and the "neuroscience of ethics". The former refers to the field of bioethics, which is responsible for addressing moral dilemmas that arise in the field of neuroscience, above all as a result of new discoveries that lead to technological innovations. The concerns of the "neuroscience of ethics" lead in a contrary direction and address the specific area of neuroscience that seeks to study moral phenomena along with all their associated categories, such as will, freedom or consciousness. Having drawn this fundamental distinction, Gracia-Calandín identifies how it could likewise be applied to the field of neuroeducation (Gracia 2018: 56–57). In response to studies by various authors, including Hardiman (2012) and Howard-Jones (2010), Gracia-Calandín distinguishes between the ethical issues that arise from the application of neuroscientific advances, discoveries and innovations in the field of educational, what may be termed an ethics of neurodcation, and neuroeducation as formally applied to the field of ethics. It is the latter distinction that is addressed in this chapter, which seeks to examine how the potential insight of neuroscience can be applied to moral education.

The particular approach taken in neuroeducation is to develop learning methods and strategies based on the empirical data provided by neuroscientific discoveries and experiments. The data is usually obtained using neuroimaging techniques and is therefore essentially observational. Thanks to these new technological developments, studies that seek to explain diverse facets of human behaviour have far surpassed other more orthodox empirical methods, such as those developed through research on behaviourism for example. As a result, a new source of scientific knowledge has opened up, one that expands the scope for developing an understanding of human nature. One likely impact of this shift within scientific practice is that observational data of the emotions will be connected to values, and in particular moral ones. Indeed, as I demonstrate in the pages that follow, emotions by virtue of their structure contribute a determinant epistemological function that aids the development a knowledge of values.

The findings made on emotions within the field of neuroscience are by no means a novel discovery. There is a long tradition in the field of moral philosophy of analysing the emotions in order to determine the role they play in moral judgements. Thus, authors of the Scottish School of Common Sense Philosophy, for example, understood the intimate link between emotion and the moral sphere; and more recently, contemporary authors such as Max Scheler have argued that values are ascribed to the conscience through phenomenological perceptions of an emotional nature. Furthermore, the tradition of uniting morality and emotion has its roots in Classical Greek thought, and above all that of Aristotle, whose ideas have provided a foundation for ideas developed by other contemporary authors, such as Martha Nussbaum. According to Nussbaum, as well as other authors, values are perceived through emotions in several ways, and as is shown below, their cognitive-evaluative structure plays a key role. Nonetheless, it is of the utmost importance to note that the recognition of the fundamental role played by the emotions in the perception of values by no means justifies a claim for emotional management being equivalent to moral education.

5.2 Moral Education Versus Emotional Management

Moral education has traditionally been understood as being closely linked to the teaching of a set of values, and even in some cases to ideology. However, of recent, the understanding of moral education as the transmission of a prescribed cluster of values has become a controversial issue. In the context of contemporary pluralistic and democratic societies, which encompass a range of world-views, ethical proposals and ways of conceptualising life, a form of education intended to inculcate or indoctrinate a particular value system is often dismissed and even proscribed in western democracies.

One factor in the contemporary shift affecting moral education is that it has become linked to the development of learning models employed in emotional management, and as a consequence moral education has even been equated with emotional management. Undoubtedly emotions and morals are closely linked but, in my view, they cannot be understood in all but synonymous manner.

One outcome of this recent development is that educational projects based on emotional management are now being implemented in schools as a means of fostering so-called social skills and improving behaviour. However, projects such as these often blur moral concerns with other types of issue that are not strictly moral ones, and they have more to do with the individual subjects' concept of self, random psychobiological factors, and their social and family environment. An example of one project that is being increasingly deployed across the USA is the SEL (Social and Emotional Learning) programme. The approach taken by this programmes proposes that the best way to tackle students' behaviour issues, which in my view is addressed in a moral sense, is by promoting social skills based on what is deemed to be the correct form of emotional management (Greenberg et al. 2003; Weissberg and O'Brien 2004; Extremera and Fernández 2005). Fernández and Extremera, in particular, argue that Mayer and Salovey's Emotional Intelligence model can obtain excellent results using a form of emotional education that encourages what are termed "social behaviours". Their model promotes: a) emotional perception, the ability to recognise and identify one's own and others' emotions; b) emotional assimilation, the ability to take emotions into consideration when making decisions, which is underpinned by a concept of the emotions as devices that enable subjects to prioritise and focus their attention, and also to separate the accidental from the essential; c) emotional comprehension, the capacity to understand complex emotions and emotional transitions thereby enabling subjects to better understand their behaviour and the causes of their actions; d) emotional regulation, a skill to discard certain emotions and to take advantage of others according to their usefulness, which is of great strategic value at a behavioural level, but training is needed to master this skill. It is this final skill, emotional regulation, that, in my view, is often assimilated into moral education. The regulation or management of emotions focuses primarily on the promotion of those emotions that are considered "positive", and in the moderation and even suppression of so-called "negative" ones.

The work of Martha Nussbaum has played a significant role in transposing the thesis of emotional management into the field of moral education, and her work is underpinned by a specific political perspective (Nussbaum 2013). Nussbaum classifies some emotions as specifically political emotions: the public emotions that foster moral attitudes and behaviours and thereby have a clear public impact. She also identifies other emotions as preventing the development of moral responses and therefore making it impossible to achieve specific goals of social justice. The idea underlying her approach is that emotions are a fundamental element of social cohesion and therefore necessary to achieve a stable concept of justice. However, Nussbaum argues that not all emotions contribute to this social concern, and some emotions such as compassion, love or sympathy are to be encouraged because they contribute to forming affective bonds that provide a solid foundation for putting into practice a theory of justice. On the other hand, emotions such as envy, disgust or shame can provoke contrary reactions and destabilise or efface a political model built on the values of a specific concept of justice. The reason for this is that the latter emotions do not foster attachment, as in the case of compassion or sympathy, and instead give rise to a sense of fragmentation. In this sense, Nussbaum argues that, in so far as it is possible, the suppression of these negative emotions is necessary for the purpose of justice and she advocates the promotion of the positive emotions.

In my view, the point of departure for her proposals is flawed by a basic error: the equation of moral education with emotional management. In drawing this distinction, I call into question Nussbaum's thesis that "positive emotions"need to be promoted and "negative ones" suppressed in order to achieve moral progress. Furthermore, in doing so I also challenge the current neuroeducational trend of promoting a form of moral education that is based exclusively on the regulation of the emotions. However, it is not my intention to deny the important contributions made by emotional management as an area of study. Instead I seek to provide a more solid foundation for its coexistence with moral education, and in my view, this should be based, firstly, on an understanding of the prominent role played by the emotions in the formation of moral judgements, and, secondly, that moral education must employ educational processes that engage with other spheres that transcend the emotions, rather than solely focusing on fostering certain emotions and suppressing others. Indeed, there is considerable risk that a focus developed merely on encouraging or stifling different emotions, will prove counter-productive as it can lead to emotional repression, a lack of critical judgement and moral blindness.

5.3 The Role of the Emotions in Moral Neuroeducation

The thesis that the emotions provide human beings with an original mode of access to the sphere of value has been upheld by a range of classical, modern and contemporary thinkers. For Scheler, a paradigmatic author within the field of the Theory of Values, emotions are primary intuitions towards the apprehension of value (Scheler 1970); the psychologists Arnold and Lazarus, founders of Appraisal Theory, argued that emotions are responses to the assessment of a situation (Arnold 1968; Lazarus 1991); and contemporary authors, such as Ronald de Sousa and Martha Nussbaum, point out that the emotions are intelligent responses to the perception of value (Sousa 1987; Nussbaum 2001).

In addition, biological research has contributed various arguments supporting Scheler's thesis that emotions are a mechanism for capturing values. Human beings are creatures endowed with a nervous system (Kiernan and Rajakumar 2013) whose fundamental function is to coordinate all the external stimuli received by the organism in order to process sensory information and thereby make appropriate responses, thereby guaranteeing human adaptation to the environment and survival. It is thanks to the nervous system that human beings, as well as other living beings, are sentient and possess a structural biological capacity to feel pleasure and pain. When the sensation of pleasure and pain transcends the purely somatic realm and becomes a psychological experience, the biological process that is initiated has a greater degree of sophistication than a purely bodily experience of pleasure or pain, and these processes are called emotions. Emotions are the psychophysical mechanism that human beings rely on, and along with various other species, it enables them to express their inner capacity of sensing pleasure and pain in conjunction with their structural intellectual skills. Emotions are configured as a device that mediates between a subject's somatic and intellectual processes and they enable interactions with the external world in order to prioritise and hierarchise experiences of it. In other words, they are configured as an invaluable survival tool for the emotional living being. Emotions are the physiological mechanism that allows human beings to interact with the external world in a particular way, one that is key for their viability and the world they create for themselves. They provide physiological and intelligent responses to certain stimuli in conjunction with a formal evaluative instance, which the being experiencing the emotions has introspective access to. Therefore, emotions are not blind physiological responses or purely somatic reflex acts, instead an intellectual component is required to trigger them.

The theory, originally developed in Classical Greek philosophy and Stoic thought, suggests that we get emotionally excited because there is something that matters to us, something we value truthfully. We know that something matters to us, or that something is valuable in an original and true way because we feel it that way. In other words, a certain event or external event (in the sense that it does not depend on our direct control) prompts a reaction in us that has a value component, which can be positive or negative, and this gives rise to an emotion. The formal structure of what constitutes an emotion is still a subject of debate today and remains to be fully clarified, but most of the aforementioned authors, including Martha Nussbaum (2001), argue that it contains four structural components.

5.3.1 The Physiological Component

Emotions have a physiological structure. When an emotion occurs, the body experiences a series of physiological processes and chemical reactions, such as the skin blushing, increased temporary heat, or depending on the emotion that arises—fear, anger, joy, etc.—increased levels of neurotransmitters, such as oxytocin or cortisol. However, study of the physiological basis of emotions, which has obvious neuroscientific implications, is not without controversy. A core issue, which was first set out in W. James's controversial *Mind* article, "What is an Emotion "(James 1884), persists today, above all in relation to the ethical dilemma and practical evaluation of the idea of human moral enhancement achieved through interventions of an exclusively physiological nature.

James ascribed a prominent role to the physiological reaction and identified it as a founding element of emotion. According to James the bodily changes follow directly from the perception of the arousing fact, and it is our sense of these changes taking place that is emotion (James 1884:189). Although common sense would state that the corporal manifestations occur once the emotion has been felt, James maintained that this sequence is incorrect and that we feel sorry because we cry, or fear because we tremble and not the contrary (James 1884:190). However, it should be noted that his conclusion of this inference makes a qualitative leap, which in my opinion makes it erroneous, as this conclusion is not necessarily derived from its initial premises.

James's theory defines emotion as the perception of these involuntary, instinctive and physiological changes. The proof he claims for this being the case is that if experience is deprived of physiological manifestations, then we do not have an emotion, but just a cold and neutral state of intellectual perception. He goes on to argue that if we imagine a strong emotion, and then try to abstract from our consciousness all the sensations characteristic of bodily symptoms, we find that nothing remains, there is no 'mental substance' that constitutes emotion (James 884:193). On this basis, the author concludes that an emotion is not the cause of physical reaction, but an emotion arises from the experience or awareness of that physical reaction.

Similar theses are encountered in the field of neuroscience, however, authors like Nussbaum argue that while the existence of a physiological basis for emotions is evident, they cannot be reduced to non-cognitive physical impulses. Although

Nussbaum's position is not totally contrary to the neuroscientific thesis on emotions, she does consider that at present there are no sufficiently convincing arguments for emotions being reduced to an awareness of purely physiological processes (James 1884: 145). Nussbaum states that emotions are embodied, bodily manifestations, which does not necessarily imply that the associated physiological processes are the cause of emotion, because the arousal of emotions requires a mental component. Indeed, emotions play a unique motivating role: despite being mental states, they produce somatic alterations that are unique and of great magnitude. In this way, Nussbaum recognises emotions as being associated with certain physiological effects and somatic alterations and, in this sense, they are processes that materialise in the physical realm. Nevertheless, this position does not necessarily imply that these physical reactions are the sole foundational component of the experience that is emotion. Given the body's considerable malleability mental states can manifest themselves in a range of very different ways, whereby it is challenging to establish, as neuroscience endeavours to do, any stable physiological patterns for emotions, and it is this fact that undermines neuroscientific claims (Nussbaum 2001:82).

5.3.2 The Physiological Component

Emotion has a cognitive structure, that is, it is not a pure physiological reaction, as it has to mediate an intellectual component, in other words it is a cognitive understanding of the event that triggers the emotion. Emotions are not mere appetites like the sensation of hunger or thirst, instead they respond to a determined object that is apprehended through a mental process.

The mental component of emotions, which is necessary to endow them with a cognitive nature, is structured around two central ideas: emotions are not mere physiological impulses and they are directed towards someone or something. In this sense, emotions contain within themselves an orientation towards an object, and within the emotion the object is confronted through an intentional description (Nussbaum 1997:94). Essentially, this turns emotion into a way of perceiving objects. And in relation to this, the idea that emotions are intimately related to certain beliefs about their object becomes crucial.

In order to argue that emotions are cognitive experiences it must be affirmed that they are *intentional*. The concept of *intentio* has its roots in Brentano's thought, although he gave no indication that emotions provide us with any cognitive content. As emotions do not inform us of characteristics that are independent of the world, and insofar as they are a subject's experiences, we can only speak of an experience of something being "correctly wanted or loved", but not in terms of having access to it or in the sense of a cognitive expansion. However, Scheler subsequently went on to argue that emotions do have a cognitive value because they allow us access to the world of values (Scheler 1970, 200). Emotions in Scheler's theory operate as first intuitions of what is valuable and what is not and, therefore, the type of cognition they provide is about the world of values. In this way, Scheler's theory may be classified as a "phenomenology of emotion"or an "intentional theory of emotion".

On the other hand, to understand the relevance of the effective linking between the emotions and a subject's beliefs, it is necessary to highlight the evident Aristotelian basis underpinning such an affirmation. In the *Rhetoric*, Aristotle sets out the idea that beliefs are an essential basis for emotion. Each type of emotion is associated with a specific family of beliefs, whereby if a person does not believe in, or stops believing in a relevant family of beliefs, he or she will not experience the emotion. Beliefs can be true or false, and this also implies that emotion can be true or false as well. Regarding the latter issue Chrysippus posed the interesting and subtle question of what is meant by a particular interpretation of the truth or falsity of an emotion. According to Chrysippus, although a belief may be false, it may successfully prompt an emotion, precisely because an emotion arises from a belief that is held to be true, regardless of whether it and the value judgements attached to it are true or not. The link between beliefs and emotions is both necessary and sufficient. However, this does not suffice to clarify the cognitive structure of emotion, as the connection between emotions and a subject's mental life of a subject must first be clarified. Likewise, the mental content under discussion must be elucidated. Having shed light on these issues it may be argued that it is clear that the emotions operate at a complex psychological level. The Stoics considered the judgements that motivate emotions to be effectively directed to the sphere of subjective desires and satisfactions, which are in turn linked to the sphere of value. Consideration of this position is examined in the next section.

5.3.3 The Evaluative Component

Emotions also have an evaluative structure, which implies that they arise as a consequence of an evaluative judgement that made about the event which the subject is affected by. As was discussed above, emotions give rise to a particular kind of cognition. It is a form of cognition that is not restricted to the cognitive information provided by perception, instead it presents perceptual information in an evaluative form. As a result, this specific cognitive information provides a type of knowledge that is introspective by nature, which also allows access to a subjective dimension as well as informing us about the world of values. Therefore, the type of cognition that emotions provide concerns information about what is valuable to the emotional subject. In other words, emotions are the means through which a subject perceives values.

Yet to define emotions as the means through which values are perceived requires a conceptual clarification. What is referred to here is nothing less than a primary access to the dimension of value and, therefore, it is not related to the act of "assessing" in the sense of "weighing" or "calculating". Essentially, deliberative reasoning operates on a different level from that of the emotions. Emotions are perceivers of value in a primary sense and, therefore, they possess the necessary characteristic identified by the Stoics - they are fresh- insofar as they constitute themselves as immediate reactions to the perception of value. In this sense, they lose intensity over the temporal course of the perception of an object. Deliberation and calculation are undertaken later and can, therefore, be guided through calm reflection, a quality the emotions lack.

In addition, the perception of values through the emotions refers to the subjective dimension of the perceiving subject, whereby it seems that the objective world of values cannot be understood unless related to the subjective world of values. Seemingly, this latter consideration adds a further layer of complication to the question of whether it is possible to draw on this theory of emotions in order to identify the form of ethical judgement within emotional-evaluative judgements themselves, insofar as ethical judgement can be considered as having a claim of necessary universality.

5.3.4 The Narrative Component

A further formal component must now be added to the physiological, cognitive and evaluative nature of the emotions, their narrative, and this renders the analysis of emotion still more complex. What it is at stake is not only a concern to affirm that emotions have a narrative dimension, which is problematic in its own right, but also to address a further level of complexity that is raised. The fact that emotions are necessarily linked to a narrative or story that develops over time, implies their consequent location and contextualisation, and thereby particularity. A number of difficulties are prompted by this view with regard to formulating statements of a universal nature and in conferring a moral status on the emotions.

Some theories of the emotions overlook memory as an essential and key component, yet it is precisely the memory that allows emotions to be triggered. Memory is an intellectual activity that enables the organism to function at a level beyond the merely instinctive. It is an essential tool for life and survival, and all human activity is organised according to memory. Any activity mediated by an intellectual component requires the prerequisite of memory for its viability. Therefore, if the cognitive nature of the emotions is to be acknowledged, it must be assumed that memory contributes to their cognitive role.

It is in regard to this issue that Nussbaum diverges from the Stoic thesis; in her classification of the emotions she notes that the Stoics omitted the past as a temporary category (Nussbaum 2001:208). Likewise, she points out, that some contemporary theories defend a cognitive conception of the emotions as detached from their past and determined by the things that affect the person in the present moment. It was specified above that emotions have a cognitive content of a particular nature, and are not mere reflexological physiological responses, yet as Nussbaum (Nussbaum 2001:210) convincingly argues this cognitive content is presented as a complex, interwoven narrative. In this regard, Nussbaum states that for an emotion to arise it necessarily presupposes a link with a succession of past states, which remain present, either consciously or unconsciously, within the subject affected by the emotion. Therefore, emotions have their roots in childhood.

The fact that the emotions necessarily possess a narrative form implies that in order to understand the concrete meaning of an emotion it is necessary to know the story that gives rise to it. It is only through analysis of an emotion's specific underlying narrative that it is possible to understand the emotion's meaning and particular scope. If, on the other hand, an emotion is studied in a biased way it will necessarily appear as an irrational impulsive force lacking in meaning. Nussbaum went on to state that ultimately emotions have a story behind them because it is not possible to reach an adequate understanding of any form of pain unless it is conceived as interwoven with a deep love story, with desires for security and protection. Likewise anger needs to be understood in the narrative context of external events (Nussbaum 2001: 206). Once the narrative structure of emotion is affirmed, its evolutionary character should also be assumed. Emotions must logically evolve, just as both a story does and the way one relates to valuable objects. Indeed, emotions connect subjects with their own past history, but, in turn, their emotions are modified according to the continued flow of their history; thereby as each subject proceeds along their biographical path they accumulate an "emotional burden", and depending how it is managed—correctly or incorrectly—it can be either counter-productive, if it blocks a "normal functioning", or a valuable adaptive tool by providing an excellent source of experience that functions at a deep level.

The fourfold structure of emotion set out here reveals the intimate connection between the emotions and the sphere of values. Therefore, there are solid reasons to argue that the emotions are a fundamental part of the moral phenomenon: they are both a confirmation and a psychobiological reflection of the moral phenomenon. Currently, the strategy proposed to contribute to the moral development of the human race is to promote emotions that foster moral behaviour. Therefore, the appropriate means should be developed to foster moral emotions. Yet the question of how to do so is not easily answered, as we are faced with a dilemma: in order to arouse certain emotions is it sufficient to merely reproduce the physiological processes that occur in different emotions through artificial procedures, such as the administration of drugs that elevate oxytocin levels? Or, taking into account the foregoing discussion of emotions as complex phenomena with a cognitive structure, would it be more appropriate to pursue intellectual practices in order to influence emotions? In the latter case, moral education would be based on emotional education and it would be facilitated by the management of the emotions. In this regard, Aristotle's view that practical wisdom consists of feeling the right passion in the right way provides a point of departure for addressing this dilemma.

In my view, the emotions are cognitive in the sense that they are informative. Or in other words their function is to serve as bridge between a subject's body and mind. They provide a somatic device that informs subjects that something is happening and puts them in a situation of "alert", a state of presentness and interest in the here and now, one that is moreover a physical, necessary, spontaneous and inevitable interest. This is the function of both the informative and motivational elements of the emotions. However, something more than factual cognition occurs when an emotion arises, which is that it must also include a belief, and this belief must be of a certain type. It must be a belief about something that is valued and appraised. It is this belief that is modifiable, and is in turn based on a previous category, which is the structure of understanding. The more open this comprehensive structure is, the more easily it can incorporate moral elements. An ever-deepening understanding is required, especially to achieve an emotional education. The more a subject is able to understand, the more difficult it is for clearly immoral attitudes such as cruelty to arise within them. Thereby, in my opinion, cruelty may be understood as a reduction of the comprehensive variables. The greater the plurality of information that is mediated by a subject's true understanding, the more difficult it is for them to be cruel and unjust, because moral education, insofar as it takes a universal form, can only occur in holistic approaches that incorporate a totalising approach.

5.4 Moral Education

The foregoing discussion allows moral education to be understood in a wholly different manner to the educational approaches developed in response to the aforementioned authors, such as Goleman, and which have since evolved intro projects such as SEL. Emotions play a prominent role in the formation of moral judgements, but not as producers of moral actions in themselves. Instead they provide a psychophysical device that enables human beings to adapt to a multiple world defined by a plurality of stimuli: emotions allow subjects to prioritise and, therefore, guide their decision-making.

In my view, those projects that conceive of moral education as being achieved by promoting certain emotions and suppressing others all commit the same basic error as the rationalist projects that have sought to extirpate the emotions from their theoretical model to ensure that it be classified as rational. The projects referred to here are above all the analytical models developed in the first half of the twentieth century. On the other hand, the Greek Stoic project that aimed to control emotions in order to achieve happiness conceded emotions a high cognitive value and considered them eminently rational.

All emotions have an epistemological value, and therefore, some qualify as positive and desirable emotions, while others are prone to suppression because their seemingly negative effects imply rendering human beings' essential nature as biased and leaving them defenceless against certain vital vicissitudes. As has been discussed, an emotion requires the emotional subject that experiences it to perceive it as information invested with a certain "value", and, therefore, the subject must be understood as being affected in some sense. Likewise, as has been noted, emotions are physiological reactions to certain events that are perceived by the emotional subject in an evaluative and interested way. The emotional perception that triggers a somatic response must be substantiated by evaluative judgements about the perceived object on which a belief resides, whether this is founded in truth or not. Otherwise, if there is no appraisal component, the emotion can be feigned and will not be consummated de facto. Emotions are not mere physiological reflex acts. In order for an emotion to arise some external information providing cognitive content, whether true or not as this is irrelevant to emotion, must first act as a mediator. What provokes the emotion is the subjective belief produced by the emotional subject with this perceived information. The belief that the emotion is not just a case of mere information processing is linked to the axiological system underpinning the subject's experience of the emotion. Emotions cannot be considered as irrational acts, at least in a cognitive sense, or in other words as devoid of propositional content. Emotions are not merely of considerable importance for the field of moral reflection, they are decisive, since in addition to establishing somatic reactions to evaluative beliefs, they are privileged "informers" that allow access to the sphere of the self and other subjects' value systems. In their ordinary life subjects live with and in accordance to their emotions. They may seek to enhance some emotions and control others, or else try to achieve a degree of emotional stability, whereby they are not overly affected by emotional change. The truth is that each subject is immersed in their particular emotional psychobiological device, which underscores how their system of values is inextricably linked to their organism. Emotions are not only a subsequent reaction to a previous belief, but they act as a regulating thermostat of the subject's axiological system, and this informs them of their own values when confronted by the events that demand action and decision-making. The unique motivational force of the emotions is, in turn, determinative of the meaning of a subject's deliberations and, therefore, of the actions they take. This is because for emotions to be triggered they must have an evaluative nature in their own structure that is not independent of themselves, but instead shapes them. For these reasons, emotions—all of them—play a prominent role in the moral sphere, and this cannot be ignored in either human moral development, or ethical decision-making.

The four elements examined above, when combined in such way that they interact can, in my view, open up a fifth dimension within the structure of the emotions: a normative level of the emotions is provided, and in this sense emotions are likewise normative. Each and every emotion offers honest information about the cluster of values that the affected subject emotionally sustains. Emotions are not chosen, instead they emerge and manifest themselves somatically as a response to an evaluative mode of cognition through which they maintain a direct and immediate relationship. It is in this sense that I argue that the emotions are normative, because they are built onto an imperative structure that in principle is not arbitrarily negotiable, but rather receives its *modus operandi* from the pre-emotional value system possessed by the subject.

Indeed, various authors, especially in both the fields of psychology, and political philosophy, including Nussbaum herself, have observed the importance of emotional education when understood in terms of emotional management and the modulation of emotions. However, the truth is that each and every emotion provides invaluable information about their subject's inner world. Emotions permit a connection with the intimate sphere of values in a truthful manner. Therefore, moral education should not focus on the regulation of the emotions because this would result in a blurring of the process of perception—which functions in a technical manner and affects the execution of actions— with the sphere of "planning", which is and must be antecedent.

It is this level of moral educational guidance that should be applied to cognitive structure, that is, the production of beliefs and practical deliberation, as it is the production of beliefs that gives rise to the emotions addressed by moral education. Therefore, despite the fact that moral education can and must involve the sphere of emotions, it should by no means solely be undertaken through emotional management, whether it be purely psychological or undertaken using neuroscientific methods. The staged processes of moral education must undergo a gradual increase of knowledge in a broad sense. Indeed, they should foster curiosity about the world and about the worlds of others, and thereby influence subjects' understanding of the structures that shape their thinking and enable them to form critical judgements. To achieve this, students of moral education must learn to listen to their emotions, and, above all, they must be educated in a holistic and comprehensive manner.

References

Ansari, D., D. Coch, and B. de Smedt. 2011. Connecting education and cognitive neuroscience: Where will the journey take us? *Educational Philosophy and Theory* 43 (1): 37–42.

Ansari, D., B. de Smedt, and R. Grabner. 2012. Neuroeducation: A critical overview of an emerging field. *Neuroethics* 5 (2): 105–117.

Arnold, M.B. 1968. *The nature of emotion: Selected readings*. Baltimore: Penguin.

Blair, J., A.A. Marsh, E. Finger, K.S. Blair, and J. Luo. 2006. Neuro-cognitive systems involved in morality. *Philosophical Explorations* 9 (1): 13–27.

Borg, J.S., C. Hynes, J. Van Horn, S. Grafton, and W. Sinnott-Armstrong. 2006. Consequences, action, and intention as factors in moral judgments: An fMRI investigation. *Journal of Cognitive Neuroscience* 18 (5): 803–817.

Campbell, S.R. 2011. Educational neuroscience: Motivations, methodology and implications. *Educational Philosophy and Theory* 43 (1): 7–16.

Chalmers, D. 1998. What is a Neural Correlate of Consciousness? In *Neural correlates of consciousness: empirical and conceptual questions*. ed. T Metzinger. published with MIT Press in 2000.

Cortina, A. 2011. *Neuroética y neuropolítica. Sugerencias para la educación moral*. Madrid: Tecnos.

Damasio, A.R. 1994. *Descartes' error: Emotion, reason, and the human brain*. New York: Avon Books.

Extremera, N. and Fernández-Berrocal, P. 2005. La Inteligencia Emocional y la educación de las emociones desde el Modelo de Mayer y Salovey. *Revista Interuniversitaria de Formación del Profesorado* 19 (3): 63–93.

Gracia-Calandin, Javier. 2018. El fin ético no naturalista de la educación. *Recerca. Revista de Pensament i Anàlisi n.* 22: 51–68.

Greene, J. 2003. From neural 'is' to moral 'ought': What are the moral implications of neuroscientific moral psychology? *Nature Reviews Neuroscience* 4: 847–850.

Greene, J.D., R.B. Sommerville, L.E. Nystrom, J.M. Darley, and J.D. Cohen. 2001. An fMRI investigation of emotional engagement in moral judgment. *Science* 293 (5537): 2105–2108.

Greene, J.D., L.E. Nystrom, A.D. Engell, J.M. Darley, and J.D. Cohen. 2004. The neural bases of cognitive conflict and control in moral judgment. *Neuron* 44 (2): 389–400.

Greenberg, M.T., R.P. Weissberg, M.U. O'Brien, J.E. Zins, L. Fredericks, H. Resnik, and M.J. y Elias, 2003. Enhancing school-based prevention and youth development through coordinated social, emotional, and academic learning. *American Psychologist* 58: 466–474.

Hardiman, Mariale, Luke Rinne, Emma Gregory, and Julia Yarmolinska. 2012. Neuroethics, Neuroeducation and classroom teaching. Where the brain sciences meet pedagogy. *Neuroethics* 5: 135–143.

Howard-Jones, Paul. 2010. *Introducing Neuroeducational research. Neuroscience, education and the brain from contexts to practice*. Londres: Routledge.

Howard-Jones, Paul, Daniel Ansari, Bert De Smet, Diana Laurillard, S. Varma, Brian Butterworth, Usa Goswami, and Michael S.C. Thomas. 2016. The principles and practices of educational neuroscience: Comment on bowers. *Psychological Review* 123 (5): 620–627.

Hume, D. 2007. *A treatise of human nature*. Oxford: Clarendon Press.

Hutcheson, F. 1971. *An essay on the nature and conduct of the passions and affections (1728)*. Hildesheim: Georg Olms.

James, W. 1884. What is an emotion? *Mind* 19: 188–204.

Kahane, G., K. Wiech, N. Shackel, M. Farias, J. Savulescu, and I. Tracey. 2012. The neural basis of intuitive and counterintuitive moral judgement. *Social, Cognitive and Affective Neuroscience* 7 (4): 393–402.

Kiernan, J., and N. Rajakumar. 2013. *Barr's the human nervous system: An anatomical viewpoint*. Philadelphia: Lippincott-Williams-and-Wilkins.

Lange, C.G., W. James, and I.A. Haupt. 1922. *The emotions, Psychology Classics*. Vol. 1. Baltimore: Williams & Wilkins Company.

Lazarus, R.S. 1991. *Emotion and adaptation*. New York: Oxford University Press.

Long, A.A., and D.N. Sedley. 1987. *The Hellenistic philosophers*. Vol. 1. Cambridge: Comabridge University Press.

Marina, José Antonio. 2012. Neurociencia y educación. *Participación Educativa* 1: 7–13.

Nussbaum, Martha. 1997a. *Poetic justice: The literary imagination and public life*. Boston: Beacon Press.

———. 1997b. *Cultivating humanity. A classical defense of reform in Liberal education*. Cambridge, MA: Harvard University Press.

———. 2001. *Upheavals of thought: The intelligence of emotions*. Cambridge: Cambridge University Press.

———. 2013. *Political emotions: Why love matters for justice*. Cambridge, MA: The Belknap Press.

Pallarés-Domínguez, Daniel. 2016. Neuroeducación en diálogo: neuromitos en el proceso de enseñanza-aprendizaje y en la educación moral. *Pensamiento* 72 (273): 941–958.

Roskies, Adina. 2002. Neuroethics for the new millennium. *Neuron* 35: 21–23.

Savulescu, J., and I. Persson. 2013. Getting moral enhancement right: The desirability of moral bioenhancement. *Bioethics* 27 (3): 124–131.

Scheler, M. 1970. *The nature of sympathy*. Trans. Peter Heath, Hamden. CN: Archon Books (first printing, London: Routledge & Kegan Press, 1954).

Schoenbaum, G., M.R. Roesch, T.A. Stalnaker, and Y.K. Takahashi. 2009. A new perspective on the role of the orbitofrontal cortex in adaptive behaviour. *Nature Reviews Neuroscience* 10: 885–892.

Sinnott-Armstrong, W. 2008. *Moral psychology, volume 1: The evolution of morality, volume 2: The cognitive science of morality, volume 3: The neuroscience of morality*. Cambridge, MA: MIT Pres.

Sousa, R.D. 1987. *The rationality of emotion*. Cambridge, MA: MIT Press.

Terbeck, S., G. Kahane, S. McTavish, J. Savulescu, P. Cowen, and M. Hewstone. 2012. Propranolol reduces implicit negative racial bias. *Psychopharmacology* 222 (3): 419–424.

Weissberg, R.P., and M.U. O'Brien. 2004. What works in school-based social and emotional learning programs for positive youth development. *The Annals of the American Academy of Political and Social Science* 591: 86–97.
Wiech, K., G. Kahane, N. Shackel, M. Farias, J. Savulescu, and I. Tracey. 2013. Cold or calculating? Reduced Activity in the Subgenual Cingulate Reflects Decreased Aversion to Harming in Counterintuitive Utilitarian Judgment. *Cognition* 126 (3): 364–372.
Winecoff, A., J. Clithero, et al. 2013. Ventromedial prefrontal cortex encodes emotional value. *Journal of Neuroscience* 33 (27): 1132–1139.

Chapter 6
Why Moral Neuroeducation Should Embrace Ethical Humour

Juan Carlos Siurana

6.1 Introduction

The central premise of this chapter is that humour has considerable potential to make a positive ethical contribution to the life of individuals and societies as a whole, and in particular with regard to healthcare. In relation to the latter I argue that the concept of "ethical humour" offers a key foundation for neuroscientific studies on humour and that it may in turn provide a basis for a "neuroethics of humour". I define humour as the ability to perceive the comic, and I propose that it reflects the behavioural habits that an individual freely takes on, especially their ethical education, which is understood here as providing an initial foundation for developing a concept of "ethical humour".

Over the course of this chapter, I argue that ethical humour is therapeutic, in the sense that it provides health benefits for individuals, and to demonstrate this I examine a key selection of these benefits in conjunction with the scientific evidence for their efficacy. In addition, attention is paid to the contrasting results of a study in which humour's contribution to a healthcare setting was considered negatively, but I argue that these results reflect the fact that the use of humour was only measured quantitatively and not qualitatively. The latter approach needs to be foregrounded as qualitative studies devoted to forms of humour, akin to what may be understood as "ethical humour" have indicated its positive influence on healthcare.

To develop this argument, a review is undertaken of neuroscientific studies devoted to the parts of the brain involved in humour and hilarity. While providing valuable insights into the cerebral processes linked to humour in my view these studies fail to take human freedom into account. In contrast, I maintain that we are free to forge our own character and, thereby, free to laugh in an ethically correct

J. C. Siurana (✉)
Universidad de Valencia, Valencia, Spain
e-mail: juan.c.siurana@uv.es

P. Calvo, J. Gracia-Calandín (eds.), *Moral Neuroeducation for a Democratic and Pluralistic Society*, https://doi.org/10.1007/978-3-030-22562-9_6

manner. Nonetheless, I do believe that these neuroscientific studies can be beneficial for ethically improving healthcare so long as they are used to define and employ the idea of "ethical humour". Finally, the conclusion to this article offers a summary of ethical humour, the virtues it promotes and its physiological foundations.

6.2 "Ethical Humour" Is the Reflection of an Ethical Character

6.2.1 Humour Is the Ability to Perceive the Comical

Charles Darwin, in *The Expression of the Emotions in Man and Animals* (1872), observed that laughter is a means of communicating with others that we feel the emotion that Rod A. Martin has termed "mirth" (Martin 2006, p. 254). At low intensity levels, mirth is expressed in the form of a weak smile. At high intensity levels, they are expressed through laughter, and if the intensity is very high, through boisterous laughter, accompanied by blushing and body movements, such as throwing the head back, rocking the body back and forth, slapping the thighs, etc. (Fry 1977). But what produces the emotion of mirth?

Mirth can have several causes (e.g., tickling), but the cause that interests us in this chapter is the perception of the comic as humour. Humour is the ability to perceive or show what is comic in the world that surrounds us, or else within us, and it causes the emotion of mirth, which in turn is expressed as a smile or laughter. This perceptual process takes place in systems located in numerous regions of the cerebral cortex, which are involved in visual and auditory perception, language comprehension, social cognition, logical reasoning, etc.

A person's greater or lesser capacity to understand funny things reflects their particular sense of humour. A refined sense of humour, a type highly relevant for an "ethics of humour", only gives rise to mirth in cases when what is perceived as comic has a special quality. The threshold for experiencing mirth can vary from one individual to another due to their general sense of humour, but it should be pointed out that it can also vary in the same individual depending on the following factors: social context, mood, state of health, degree of fatigue and ingestion of alcohol or psychoactive drugs. Another factor that should also be added, which is especially relevant, is the individual's ethical education.

6.2.2 Humour Reflects an Individual's Ethical Education

Smiling and laughter are expressions of the mirth caused by the perception of a physical or psychological stimulation as funny. We can simulate such stimulations and feign that we feel the same mirth, but this is not the case with authentic smiling and laughter, which only arise when something has prompted true mirth. It is usually stated that authentic smiling and laughter are involuntary. However, I would argue that this view is based on an inadequate concept of the idea of "will". What the above statement means is that we cannot avoid laughing at what we find funny, but this is highly questionable, given that we often decide to, and successfully contain our laughter when we consider it inappropriate. In a deeper sense, what this statement also seeks to say is that we cannot avoid revealing who we are, and in this sense, laughter is one of the most reliable indicators of an individual's values. However, I propose that we can in fact morally educate human beings to laugh at what is most ethically valuable, and to be incapable of laughing at what is immoral. We would, therefore, teach people to contain inappropriate laughter, and also to not perceive the humour in certain acts or jokes. We can also teach the ability to recognise and display an ethically positive form of comedy. Thereby, it may be confirmed that laughter is voluntary and a reflection of our character, which is constructed through the way we act upon our will.

To establish how to teach smiling and laughter, it is pertinent to recall the Classical philosophical theories that sought to understand the causes of this phenomenon. Scholars of the philosophy of humour agree that, over the course of the history of philosophy, at least three important theories on humour have been developed.

(Morreall 1987):

(a) *Superiority Theory.* This is the oldest of the three theories. It claims that we judge a situation to be comical when we see someone fail, while we believe we would be more likely to succeed in such a situation. We laugh at defects in others. In this context, laughter expresses a feeling of superiority (Plato 1978; Hobbes 1839).
(b) *Incongruity theory.* According to this theory, we laugh when what happens undergoes an unexpected and incongruous change that suddenly surprises us. In this case humour lies in the distortion of how things normally function, a step beyond the normal (Kant 1892).
(c) *Release of tension theory*. Finally, this theory states that humans act under pressure, with a sense of struggle and tension, all of which excites the nervous system. Laughter provides relief from nervous or mental tension and a sense of restoring our inner equilibrium (Freud 1963, 1987).

In an earlier study (Siurana 2012), I undertook a survey of a series of key ideas regarding humour as developed over the course of the history of philosophy, and this revealed clear links between the theories that have been developed.

With regard to the three theories set out here, the one with the clearest negative moral implications is, undoubtedly, the superiority theory. Laughing at the defects of others can be immoral in many cases. However, humour can have many positive effects if practised ethically (Siurana 2015). To demonstrate one facet of these positive effects attention is now turned to how "ethical humour" can contribute to improving health.

6.3 "Ethical Humour" Is "Therapeutic Humour"

One of the primary concerns of authors who address the relationship of humour to health is the question of defining a concept of humour that is truly healthy, or, as is often said, a "therapeutic humour". One of the most relevant books on this subject is Begoña Carbelo's *El humor en la relación con el paciente* (Humour and patient relationships) (Carbelo 2008). Carbelo defines "therapeutic humour" as that which "facilitates communication and empathy between patient and practitioner, and creates an appropriate context for expressing emotions, questions and fears" (Carbelo 2008, p. 31). Carbelo explains humour's contribution to health in the following terms: "humour is capable of providing an environment of healthy laughter, which helps manage stress and mediate adaptation, yet without causing offence, and without undervaluing experiences" (Carbelo 2008, p. 39).

Carbelo's definition has important ethical implications, given that the author argues that there are proper and improper types of humour. However, in my view not all humour is therapeutic, only "ethical humour" is. Therefore, my concern is to identify what makes humour ethical, as this will in turn provide a concept with considerable potential for improving both human ethics and human health. As a first step towards this goal, attention is now paid to some of the health benefits linked to "therapeutic humour".

6.3.1 "Therapeutic Humour" Stimulates the Immune System and Combats Stress

The need to address multiple issues with a sense of urgency causes physiological changes. For example: the heart beats faster and blood vessels contract, which increases blood pressure; the breathing rate accelerates providing our body with more oxygen; blood is withdrawn from the surface of the skin and sent to the muscles and the brain in order to react to a possible danger; digestion is disrupted and the pupils dilate to increase visual acuity. In addition, this defensive reaction leads to adrenaline and corticosteroids being circulated through the blood, which are only eliminated through hyperactivity. However, if this does not occur, the body suffers an onset of stress, muscle tension and even an immunosuppressive response, which is a breeding ground for certain psychosomatic diseases.

Berk et al. (1989) performed the following experiment: they studied ten healthy males; five of the participants watched a comedy video, while five control participants did not. Blood samples were obtained with the following results: the adrenaline levels of the experimental group were significantly lower than those of the control group at all times. The experiment also indicated that laughter reduced serum levels of cortisol, DOPAC, epinephrine and growth hormone. During episodes of stress, increased levels of cortisol are produced, which has an immunosuppressive effect. Laughing, in contrast, stimulates the immune system and allows it to overcome the immunosuppressive effects of stress.

6.3.2 "Therapeutic Humour" Stimulates the Brain to Secrete Endorphins, Which Have an Analgesic Effect

Research has highlighted the brain's involvement in emotional states: endorphins, which have a molecular composition that is similar to that of morphine, are secreted by the brain have a powerful analgesic effect. Ulcers, eczema, migraines, high blood pressure, increased blood glucose levels and depression appear after periods of uncertainty, disappointment, disillusionment and despair. In contrast, positive emotions stimulate the secretion of endorphins and anandamides that protect the immune system. A demonstration of this has been provided by Yoshino et al. (1996, p. 793–794), who compared the impact of humorous stories on a group of 26 patients with rheumatoid arthritis and a second group of 31 healthy individuals. An improved mood and lower degree of pain were identified in the group suffering rheumatoid arthritis.

A person's sense of humour has also been linked to an ability to stimulate endorphin production. For example, Weeks (2000) highlighted the importance of the production of endorphins in the pituitary gland, and also listed a series of activities that both increase and decrease endorphin activity, which includes laughter and smiling. Foley et al. (2002, p. 184) have also demonstrated that laughing is positively linked to optimism and self-appreciation. Subsequently, Begoña Carbelo summarised these various findings declaring, "Humour, as an attitude towards life, as a means for coping with adverse conditions, favours the synthesis of endorphins, neutralises anxiety and provides a biochemical balance for the body" (Carbelo 2008, p. 93).

6.3.3 Other Benefits of "Therapeutic Humour" in a Healthcare Context

(a) *Laughter reduces the distance between people*

Various studies by psychologists and sociologists have reached the conclusion that laughter is always accompanied by an element of reciprocity and that its primary objective is to reduce interpersonal distance (Burns 1953; Fox 1953; Coser 1959). Doctors can address their patients in a way that prompts a smile, for example: "Congratulations for coming to see me, given the weather". Phrases such as this break the ice, greet the patient, make them feel welcome and convey the doctor's wish that they should feel at ease. They also ease anxiety and improve the patient's mood, thereby helping provide quality care within which close human relationships take precedence.

(b) *Humour promotes cohesion within a work group*

According to Wooten (1997), laughter is a reliable indicator of the mental health of a team's members. If no sound akin to laughter is to be heard within a professional unit, it may be that health concerns will arise. Ideally, one should create a workspace that brims with healthy, shared laughter. Humour allows us to address situations of conflict in a more positive manner, and it both promotes group cohesion and fosters feelings of belonging (Strickland 1993). Furthermore, people want to work in an environment that values creativity and imagination.

(c) *Humour allows us to see the funny side of things*

Bokun (1989) linked laughter to a reduction of fear and postulated that when laughter occurs fears vanish. It may be argued that the explanation for this is the fact that one cannot experience two contrary emotions, such as happiness and fear, at the same time. There are two facets to human reality, one tragic and one comic, and humour is the biochemical and emotional mechanism that enables us to see the comic side of life.

(d) *Laughter improves the feeling of well-being*

Today we also know that laughter activates the cerebral cortex (Champman and Foot 1996) and that humour relieves states of anxiety and depression, as well as also accentuating a feeling of well-being. Black (1984) and Fry (1992) reported, in their respective studies, that laughter re-established homoeostasis, stabilised blood pressure, improved breathing and oxygenated the blood.

(e) *Humour increases self-esteem*

Porterfield et al. (1988) highlighted the role of humour as an important inducer of self-esteem and self-awareness, which are the foundation of a positive idea and image of oneself. Humour achieves this because it conveys a sense of appreciation, and a feeling of being better appreciated leads to individuals developing a more positive concept of themselves, which in part, is due to their increased self-esteem.

(f) *Humour helps express the deepest fears*

In the tensest conditions, humour is a resource that helps express the deepest of fears. If medical practitioners learn to recognise these comic expressions of fear and anxiety, they can obtain information about how their patients live with their health conditions (Vaillant and Vaillant 1990).

(g) *Humour distracts us from what threatens us*

In an experiment by Yovetich et al. (1990), two groups of students were told they would receive a powerful shock at the end of a period of time. During that time period, one of the groups watched a comedy film while the other group did not. The group who watched the film and thereby experienced humour displayed lower levels of stress than the members of the other group, and the experience also distracted them from the pending shock.

(h) *Humour distracts from the symptoms of pain*

Pain is a multidimensional event and an individual experience, one that influences an individual's quality of life. Pain is accompanied by an affective reaction, which usually displays anxious or depressive characteristics. In their review of articles on this subject, Matz and Brown highlighted that laughter has a considerable ability to distract from symptoms of pain (1998). For examples, McCaffery developed an experiment to demonstrate the efficacy of using humour as the primary method of distraction for patients who could not be distracted from their pain by other means. The patients listened to a recording of funny comments, which provided them with 10 min of relief. Positive results have been achieved by a number of other initiatives in humour therapy undertaken in the form of scheduled visits to cancer and other chronic disease units where patients suffer from pain. These initiatives employ comic theatre productions, humour books, games, etc. The ultimate goal is to entertain patients and distract their attention from the symptoms of pain.

6.4 The Risk of Not Working with a Concept of "Ethical Humour" When Assessing the Effects of Humour in a Healthcare Setting

A study of healthcare workers who work with AIDS and cancer patients suggested that the use of humour as a coping strategy can actually have negative consequences rather than positive ones (Dorz et al. 2003). The coping styles of 528 physicians and nurses from 20 hospitals in Northern Italy were assessed according to a measure known as COPE (Coping Orientation for Problem Experiences) (Carver et al. 1989). In addition, the participants recorded their levels of anxiety, depression and emotional distress. Surprisingly, the findings revealed that a higher frequency of humour being used as a coping strategy was associated with emotional depletion and increased feelings of depersonalisation. In my view this finding is due to the fact that the study did not address the specific sense of "ethical humour". Other studies that have analysed the qualitative dimension of humour as a coping style in medical practice, and these have concluded that it is an aggressive type of humour that is disruptive to healthcare.

Joan Sayre (2001) performed qualitative research on the use of humour by the staff of a psychiatric unit. She identified two categories of humour in use: a) a "funny" fairly benevolent type (incongruous occurrences, self-denigrating humour) and b) another "sarcastic", more aggressive type (disparaging, malicious humour). It was the latter, sarcastic humour, that was more commonly used by the staff, and the majority of the humour was dedicated to mocking the behaviour of patients, albeit when out of earshot. Sayre suggested that these more aggressive forms of humour appeared to promote negative and cynical attitudes towards the patients, which in fact could not only prove to have harmful as opposed to therapeutic results, but also give rise to ethical problems.

In addition, a further problem arises with regard to the scale commonly used for measuring coping humour (the COPE scale), which is that it does not distinguish between the potential benefits of participatory and self-assertive humour, and aggressive and potentially harmful and/or self-destructive styles of humour. However, this is measured by the HSQ scale (Humour Styles Questionnaire) proposed by R. A. Martin et al. (2003).

One of the most important concerns addressed in this chapter is to confirm that it is "ethical humour" that is a genuinely therapeutic humour. We therefore need to define, as precisely as possible, what is meant by "ethical humour", otherwise, contradictory conclusions about the effects of humour in healthcare will arise, such as those discussed with regard to humour as a coping style. Furthermore, a second concern discussed in the pages that follow is that the neuroscientific studies being undertaken on humour are failing to take advantage of a major opportunity for improving healthcare, and this is due to their inadequate definition of the idea of "ethical humour". The next section provides examples of this failing within neuroscientific research on humour, and this is in turn analysed from an ethical perspective.

6.5 The Process of the Perception and Understanding of Humour in the Brain

Neuroscientific knowledge on humour has been developed by studying brain injuries, and the use of electroencephalogram and neuroimaging technology. The majority of the results discussed here have been recorded by Rod A. Martin (2006). Regrettably, Martin's research cannot be discussed in detail, and instead I highlight a number of conclusions extracted from his research, which I consider relevant for developing a "neuroethics of humour" (Siurana 2015).

6.5.1 Patients with Right Hemisphere Damage (RHD) Suffer from Personality Changes, and a Tendency to Laughing Inappropriately or Offensively

Evidence from patients with RHD caused by apoplexy or other brain lesions suggests that the brain's right hemisphere plays an important role in processing humour. These patients have normal language skills but frequently undergo pronounced personality changes, develop socially inappropriate behaviour, and make humorous comments that are often rude and/or offensive, as well as laughing inappropriately (Brownell and Gardner 1988).

6.5.2 Patients with Left Hemisphere Damage (LHD) Do Not Suffer from Personality Changes and Understand Humour Better than Patients with RHD

Patients with LHD do not show signs of personality changes, or inappropriate social behaviour. However, they are frequently aphasic (due to language impairment, because language functions are located in the left hemisphere in right- handed individuals), but they do have a normal level of social comprehension. Significantly, they are able to extract the core message of a story.

Amy Bihrle et al. compared the ability to understand humour in patients with RHD and LHD (Bihrle et al. 1986) by using comic strips with no words. The participants were presented with the first three frames of a comic strip and asked to select one of two other frames that provided the funniest ending. The patients with RHD performed significantly worse than those with LHD when it came to selecting the correct ending for the joke, which suggests that the brain's right hemisphere plays a special role in understanding humour.

6.5.3 The Left Hemisphere Perceives an Incongruity, and the Right Hemisphere Resolves It in a Socially Appropriate Manner

Other experiments have suggested that the brain's left hemisphere plays a role in the perception of incongruity, while the right is important for providing a coherent sense for that incongruity (i.e., in resolving it) within a social context (Bihrle et al. 1988; Gillikin and Derks 1991; McGhee 1983).

6.5.4 Within the Right Hemisphere, the Right Frontal Lobe Is the Most Important Element for Understanding Humour

A study by Prathiba Shammi and Donald Stuss from the University of Toronto (1999) indicated that it is the right frontal lobe in particular that is the most important part of the brain for understanding humour. Shammi and Stuss found deficits in humour comprehension solely in patients with damage to the right frontal lobe. It seems that this lobe plays a key role in the integration of cognition and emotion, due to its connections with the limbic system, as well as with other regions of the cortex.

6.5.5 Both Hemispheres of the Brain Work Together During Humour and Mirth

Researchers have studied healthy individuals to chart the areas of the brain that participate in humour. They used electroencephalogram (EEG) technology to measure the electrical activity of the brain. Sven Svebak (1982) at the University of Bergen, Norway, measured the discordant alpha wave activity that occurred in subjects' right and left occipital lobes while they watched a comedy. Those subjects who laughed while watching the film showed lower discordant right-left alpha wave activity as opposed to those who did not laugh, which suggests that activity between the two hemispheres is coordinated during the experience of mirth. As a result of studies such as Svebak's it appears that both hemispheres of the brain work together during humour and mirth.

6.5.6 The Incongruity that Causes Laughter Is Manifested in the Brain by a Negative Polarisation Following a Positive Polarisation

In another EEG study on humour, Peter Derks and his colleagues at NASA examined ERPs (event-related potentials) associated with the comprehension and appreciation of jokes (Derks et al. 1997). ERPs are extreme disorders in the activity of positively or negatively polarised brain waves, which are produced during very brief intervals after an event and that have been found to indicate different types of information processing.

Using EEG21 electrodes at various locations on the scalp, the brain activity of the participants was monitored while they were shown a series of verbal jokes on a computer screen. Electromyographic recordings (EMG) were also taken of the zygomatic facial muscle to detect the presence or absence of smiling and laughter, in order to indicate whether each joke was funny for the subject.

The results showed that all the jokes, regardless of whether smiling or laughter was recorded, produced an increase in positive brain wave polarisation with a peak amplitude of approximately 300 milliseconds (P300) after the presentation of the ending phrase. In the case of the jokes associated with activity in the zygomatic muscle, this increase in positive brain wave polarisation was followed by a negative polarisation with a peak amplitude of approximately 400 milliseconds (N400). This N400 wave did not occur after the jokes that were not considered funny. Previous studies have shown that P300 waves indicate cognitive categorisation activity (activation of a scheme for making sense of information), while the N400 waves indicate the termination of this process and the search for an alternative regimen due to the detection of incongruity. Therefore, the signature of humour seems to be the activation of two or more incompatible schemes. The study found similar activity levels in both hemispheres of the brain, which reaffirms the idea that both hemispheres participate in the brain's processing of humour. Subsequently, these results were reproduced, and the experiment undertaken in greater depth by Seana Coulson and Marta Kutas (2001).

6.5.7 *The Brain's Left Hemisphere Performs the Cognitive Processing of Humour and the Mesolimbic Reward Network Develops the Corresponding Pleasant Emotion*

Modern progress in neuroimaging techniques such as fMRI have enabled researchers to study the regions of the brain that are involved in a wide assortment of psychological processes in normal individuals. The fMRI technique uses high-powered, rapidly oscillating magnetic fields to scan the brain and detect small changes in the levels of blood oxygenation (which are indicative of changes in neuronal activity) in specific regions of the brain. Several studies have used this method to examine humour.

An fMRI study performed at the University of Stanford (Mobbs et al. 2003), consisted of the following: While being scanned, the participants looked at, in random order, 42 funny frames and 42 unfunny control frames. Several of the brain regions that displayed a higher degree of activity when subjects were presented with the funny frames were located in the left hemisphere of the cerebral cortex, which implies they were involved in the cognitive processing of humorous information. Included among these were the following zones:

1. an area at the confluence of the left temporal and occipital lobes, which the authors suggest plays an important role in the perception of incongruous or surprising elements in humour;
2. a section of the left frontal lobe that includes Broca's area, which participates in the semantic process and in integrating language and long-term memory, and is also considered to be important for perceiving consistency and resolving incongruity;
3. the supplementary motor area of the left frontal lobe, which is linked to the expressive motor aspects of smiling and laughing.

This study detected that the funny frames also produced a greater activation in several subcortical regions including the anterior thalamus, ventral striatum, nucleus accumbens, ventral tegmental area, hypothalamus and amygdala. These regions form the core of the so-called mesolimbic reward network, which is a system that uses dopamine as its principal neurotransmitter and participates in a variety of activities that result in emotionally pleasant rewards. There was a significant positive correlation between the assessments of comedy in certain frames and the degree of activation of the nucleus accumbens, which suggests that this structure is related to the pleasant emotions associated with humour. These results have since been reproduced in subsequent studies (Azim et al. 2005; Mobbs et al. 2005).

6.5.8 *Laughter Involves Numerous Regions of the Brain*

Rod A. Martin concluded that, although the studies of humour in patients with brain lesions seem to suggest a particularly important role of the right hemisphere, brain imaging research (like the aforementioned EEG studies) indicates that humour involves coordinated activities of many regions in both of the brain's hemispheres (Martin 2006).

Laughter is a complex activity that involves cognition, emotion and motor response, which requires the coordinated activation of numerous regions of the brain, including the cerebral cortex, the limbic system and the brainstem.

6.6 The Question of Freedom: A Key Issue for a Neuroethics of Humour

6.6.1 *Are We Free to Laugh?*

To construct a "neuroethics of humour", the location for human freedom must first be established using neuroscientific discoveries. If laughter were completely involuntary there would be no sense in talking about our responsibility for our laughter, but what do the contributions of neuroscience reveal about the wilfulness of our laughter? To answer this question once more recourse is made to studies on brain lesions, and the use of brain imaging and the monitoring of the brain's electrical activity.

6.6.2 *The Emotion of Hilarity Causes Involuntary Laughter*

A number of patients, who have experienced apoplexy or other forms of brain damage that disables voluntary movement of their facial muscles, are however able to smile and laugh normally when they find something funny, or in other words when they perceive something as funny and experience the emotion of mirth.

6.6.3 *We Can Laugh Voluntarily Without Feeling the Emotion of Mirth*

On the other hand, some patients with lesions in regions of the subcortical nuclei, as well as other regions such as the basal ganglia, for example those with Parkinson's disease, are incapable of showing spontaneous facial emotional expressions when they experience subjective feelings of enjoyment, but they can laugh voluntarily if they wish to (Wild et al. 2003).

6.6.4 *Involuntary Emotional Smiling Activates the Limbic System and Voluntary Unemotional Smiling Activates Areas of the Frontal Cortex*

Iwase et al. (2002) published the results of a study using positron emission tomography (PET), another brain imaging technique. They scanned the brains of healthy participants when smiling both spontaneously, in response to comedies, or voluntarily while watching non-comedic films. The results showed the following:

(a) *Involuntary emotional smiling* resulted in a greater activation of the areas of the cortex involved in the processing and integration of visual information, as well as those areas of the cortex related to the limbic system and involved in emotional rewards.
(b) The *non-emotional voluntary facial movements* that imitate smiling lead to an increased activation of areas of the frontal cortex involved in voluntary facial movement.

6.6.5 *The Electrical Stimulation of an Area in the Left Frontal Lobe of the Cortex Resulted in the Emotion of Mirth, But Without a Corresponding Cognitive Component*

Fried et al. (1998) published a report on a 16-year-old patient who began laughing on a regular basis whenever a small region of the supplementary motor area of her brain located in the left frontal lobe of the cortex was stimulated. The laughter was accompanied by subjective feelings of happiness and hilarity in the patient. Of further interest is the fact that each time the patient laughed due to the electrical stimulation she attributed her laughter to various stimuli in her environment. For example, the patient reported that she laughed due to the funny appearance of a painting of a horse that she had looked at. Thereby, it is clear that the cognitive components of humour can be dissociated from the emotional and motor components of mirth and laughter.

6.6.6 *laughter is prompted by perceptions of humorous incongruity or a recollection of humorous memories, which activates sites in the brain that induce the emotion of mirth. However, other parts of the brain can participate in the inhibition of inappropriate laughter.*

Studies (Arroyo et al. 1993; MacLean 1987; Parvizi et al. 2001; Wild et al. 2003) have shown that the structures and systems that support laughter and mirth are found throughout the brain and include regions of the neocortex, the basal ganglia, the diencephalon, the limbic system and the brainstem. Parvizi et al. (2001) have concluded that normal emotional laughter is initiated by *humorous perceptions of incongruity, or the recollection of humorous memories*, involving association areas in the cerebral cortex. The latter activate several *emotion induction* sites located in the telencephalon (cerebral cortex and limbic system), which participate in "connecting" the emotion of mirth; these sites probably include areas of the prefrontal ventromedial cortex, basal temporal lobe, anterior cingulate, amygdala and ventral striatum (part of the basal ganglia).

When activated, the emotion induction sites operate on *emotion effector* sites (expressors), including: motor and premotor areas of the cerebral cortex, which initiates facial and body movements; the hypothalamus, in order to direct autonomous responses for our body's internal functioning, such as increased heart rate and blushing; the thalamus; periaqueductal grey matter; the reticular formation; the cranial nerve nuclei to control facial, laryngeal and respiratory actions; and parts of the brainstem, all of which are involved in smiling and laughter as expressions of mirth.

Most authors agree that there is probably a common final pathway for laughter located in the *brainstem (possibly in the dorsal area of the protuberance),* which coordinates the respiratory, laryngeal and facial components of laughter (Wild et al. 2003). Laughter is triggered in this location by the input of data from several effector sites, and outgoing signals are sent from here up to the cranial nerves in order to trigger the relevant muscles.

The excitatory triggering input data that gives rise to laughter are accompanied by inhibitory signals that reach the brainstem from several higher centres of the brain, such as the cerebellum, and these serve to inhibit inappropriate laughter.

6.6.7 The cerebellum can play a role in modulating the intensity and duration of laughter

Parvizi et al. (2001) have also considered the hypothesis of the cerebellum's possible role in modulating the intensity and duration of laughter. According to this view, the cerebellum receives information about the current socio-emotional context from structures in the cortex and telencephalon and this information is then relayed to various effector sites.

In this way, laughter can be inhibited or amplified, depending on its appropriateness for the emotional and social situation, for example, depending on whether one is at a party, or a funeral. However, when apoplexy or some other disease causes lesions in specific regions of the cerebellum or relevant structures and routes that lead in or out of the cerebellum, this modulation does not take place, causing pathological laughter in inappropriate social and emotional contexts (Parvizi et al. 2001).

6.6.8 *The reaction to tickling reinforces the idea that the cerebellum modulates laughter depending on contextual information.*

When we are tickled, we respond by laughing, but it is impossible for an individual to tickle themselves. Why does this happen? A study using fMRI examined the differences in brain activity between participants who attempted to tickle their own hands and participants whose hands were tickled by a researcher (Blakemore et al. 1998). The results showed less activity in the cerebellum when the tickling was self-induced than when the tickling was induced by someone else, suggesting that the differentiation process may be located in this part of the brain's structure.

6.6.9 *Listening to others laugh activates parts of the brain that contribute to us experiencing the emotion of mirth*

A number of fMRI studies have undertaken research on the brain areas that are activated in response to listening to the sound of laughter. Gervais and Wilson (2005) have suggested that there is a response centre for laughter that consists of specialised mirror neurons that are activated when a person observes another individual performing a certain behaviour. An fMRI study by Kerstin Sander and Henning Scheich (2001) detected that listening to either laughter or crying triggers a powerful activation of the amygdala, which is part of the limbic system and an important centre for processing emotions activated by humour.

6.7 Malicious Laughter and the Virtues of Humour

7.1 Malicious laughter reflects the unethical dimensions of the character of the person that laughs

De Sousa, one of the few authors who has written about the possibility of developing what he terms an "ethics of laughter", has focused his research on a type of laughter that is susceptible to moral condemnation: phthonic laughter or "malicious laughter". De Sousa invokes the Greek word *phthonos* used by Plato (1978) to mean something akin to "malicious envy", and which also alludes to carrying out bad actions, as well as the ambiguity between the senses of identification and rejection that characterise jealousy. Plato used this term to refer to the types of laughter typically witnessed in a ridiculous spectacle. Malicious ridicule is above all directed against our enemies and assumes that those who laugh share our animosity towards those who are laughed at.

De Sousa states that when one tells a sexist or racist joke, one assumes that the listener shares these attitudes. The phthonic element of humour requires listeners' support, and only makes us laugh when we share the speaker's attitude. A key aspect of this humour, which is highlighted by de Sousa, is that we cannot find something amusing by imagining that we share the presupposed attitudes; instead we must genuinely share them. De Sousa goes on to state that one cannot think with ethically incorrect values unless we have those values.

De Sousa also declares that thinking something is inherently amusing is an attitude and laughing at a phthonic joke reveals that we have attitudes that are actually less than ethical. To summarise, the phthonic joke depends on attitudes, and "*attitudes are beliefs that one cannot adopt hypothetically*" (De Sousa 1987, p. 241). According to de Sousa, this provides an insight into the category of emotional attitudes: they are by nature non-hypothetical. Those who laugh when it is ethically incorrect, laughs because their values are ethically incorrect, which de Sousa expressed as follows: "The 'unethical' (…) implies an *incorrect assessment of reality*" (De Sousa 1987, p. 244).

7.2 The virtues of humour

Among the authors who have written on the ethics of humour is John Morreall, whose book, *Comic Relief. A Comprehensive Philosophy of Humor*, is of particular relevance as it contains two chapters devoted to the ethics of humour (Morreall 2009, pp. 90–124). Among the most widely researched and relevant issues to the ethics of humour are racist and sexist jokes. For Morreall, although consideration of these types of joke is clearly important, it is still a small part of what has been written on the moral dimension of humour in recent decades. In his book, he distinguishes between two types of *ethics of humour*: the *negative* (where he defends the importance of limiting humour because it promotes vices) and the *positive* (where he describes the moral virtues that can be promoted through humour).

I will now demonstrate the positive type of ethics of humour by setting out the virtues that can be fostered by it. According to Aristotle, it is important for an individual to make time over the course of their life for leisure and entertainment. However, some individuals pursue entertainment to excess, and Aristotle termed them "vulgar buffoons". However, between buffoonery and complete seriousness there is a middle ground where we find happiness, and this is achieved by practising humour at the proper time and place and to the appropriate degree. The latter is termed *eutrapelia* and may be understood as related to the idea of a friendly joke. In the Aristotle's *Nicomachean Ethics eutrapelia* is identified as a virtue (Aristotle 1941).

In his *Summa Theologica*, Thomas Aquinas addressed both humour and play, and took up a position resembling that of Aristotle's: human beings require occasional rest from serious activity; humour and other forms of play offer us this rest. Thomas Aquinas called individuals who have the moral virtue associated with play and humour *eutrapelos*, by which he referred to pleasant individuals with a joyful mental attitude that gives their words and deeds a happy twist (Aquinas 1972).

Morreall goes on to develop Aristotle and Thomas Aquinas's proposal that humour can be virtuous. It should be noted that both thinkers understood virtue as a type of excellence in an individual and distinguished between intellectual and moral virtues. In response to this Morreall states that humour can promote both types of virtues, and with regard to moral or ethical virtues he argues that humour promotes the following:

(a) *Patience*. By attending to things with a sense of humour, we do not expect them to occur at the rate that would give us the most pleasure.
(b) *Tolerance*. Humour gives rise to an opening up of the mind and, with regard to social interactions, a willingness to see things differently that allows us to better understand others, what they think and how they act. Thus, humour can help reduce social friction. Sammy Basu has studied how humour promotes religious tolerance (Basu 1999). The following joke can, perhaps, serve in a mundane manner: "Have you noticed that, when you drive, anyone going slower than you is a 'Sunday driver' and anyone driving faster than you is driving 'like a maniac'?"
(c) *Kindness*. Telling jokes allows others to relax and not feel threatened. An individual who is corrected gracefully is more likely to listen to the message and act accordingly. In contrast, when people threaten others with legal action when they feel unjustly treated it puts people on the defensive, and makes them unreasonable and even hostile.
(d) *Humility*. When one laughs at oneself, situations of conflict can be averted, and the individual is appreciated by others.
(e) *Perseverance*. When humility is combined with patience and we see ourselves from a comic perspective, it is less likely that we feel overcome by frustration. Thomas Edison attempted more than 10,000 combinations of materials in order to invent the light bulb. When asked whether he felt sad for all the failed attempts, he responded, "No, I simply learned thousands of methods of how not to make a light bulb".
(f) *Courage*. The virtue of courage occurs when perseverance operates under dangerous conditions. Some insightful examples occurred during the Holocaust, as Steve Lipman recorded in his book *Laughter in Hell* (1991). For example, in the ghettos, Hitler's *Mein Kampf* (My Struggle) was referred to as *Mein Krampf* (My Cramp).

Morreall concluded that humour promotes numerous virtues, and individuals who lack the virtuous type of humour are at serious risk of becoming mere pedants. A simple analysis of the content of virtuous humour reveals how the virtues it promotes are also very relevant for delivering and improving healthcare, and in relation to this it is of considerable interest to investigate the physiological basis of the "ethical humour" that promotes these virtues.

6.8 The Physiological Basis of Humour as an Enhancer of Virtues

The type of humour examined by Morreall (1997, pp. 59–90) is that which causes individuals to laugh at their own experiences. We are the only animals who enjoy incongruity. Any animal feels negative emotions if an expectation is not met, but we can laugh. In understanding why something happens, we can identify a connection between the physiological basis of humour and the development of different virtues.

8.1 The physiological basis of the connection between humour and the intellectual virtues

The intellectual virtues refer to the appropriate exercise of thought. For Aristotle, the intellectual virtues are technology, science, prudence, wisdom and intelligence. Morreall believes that there is a physiological basis for the connection between humour and the development of these virtues of thought. The human capacity to enjoy incongruity is a function of our rationality: we can think about our experiences abstractly and objectively and thus react to them in an amusing manner. By finding a situation funny, we can transcend practical concerns and enjoy its incongruity. Rather than fighting or fleeing, as is the case for animals, we can enjoy the experience.

By promoting rational methods of thinking, humour also allows us to develop intellectual virtues such as the opening up of the mind. Individuals who are not open to new information and new ways of thinking are not only perceived as lacking in humour (Dixon et al. 1986), but also requiring more time to recognise something as humorous (Miller and Bacon 1971). Humour also engages with creative thinking, which explains why brainstorming activities often begin with humorous exercises. As Edward De Bono commented, humour is, by far, the most significant activity of the human brain. Humour shows how perceptions set up in one way can suddenly be reconfigured in another way. This is the essence of creativity (De Bono 1993, p. 8). Humour is also valuable in education for its ability to promote critical thinking (Stopsky 1992).

8.2 The physiological basis of the connection between humour and ethical virtues.

Returning to Aristotle, it may be stated that ethical virtues provide a frame of reference for the development of the human character through social customs. A person's character is forged in society. Some ethical virtues belong to self-control (strength, temperance), while others belong to human relationships (generosity, kindness, sincerity, magnanimity, justice, etc.). Yet, what physiological basis connects humour with the development of these virtues? According to Morreall, not only intellectual virtues but also ethical virtues are promoted by humour, and once more attention must be paid to the opposition between negative emotions and humour. In almost all situations in which we tend to respond with anger or fear, the morally virtuous thing to do involves overcoming the negative emotion. When our actions are the immediate result of negative emotions, we are not acting in a completely rational way. When we are overcome by anger, we can hurt or kill someone; when we are overcome by fear, we may protect ourselves at considerable cost to others. By reducing negative emotions, humorous entertainment allows us to maintain self-control and thereby act in a more rational and responsible manner.

In addition, responding to a negative situation with laughter, rather than with negative emotions, is morally important in another sense. When we laugh at an incongruity that involves our own failures and deficiencies, we see ourselves more objectively than when we do so in the light of negative emotions. By laughing at ourselves, we see ourselves as others see us, instead of only seeing ourselves from within. This ability is essential for the development of a moral perspective. As Robert C. Roberts states, a sense of humour about one's *own* foibles provides a capacity of character-transcendence; character-transcendence is basic to the very concept of a moral virtue (Roberts 1988). An individual who can only think in terms of "here", "now" and "me" is either infantile or sociopathic and, in any case, is highly distanced from morality. Only if I am able to see myself as a human in a world of other humans, and also understand what it would be like to be someone else, will I become capable of morality.

As ethics wants to transcend the perspective of the "here", "now", "me", it asks us to avoid anger, fear and other emotions focused on "me". Given that humour reduces these emotions, it plays a considerable role in enhancing ethical virtues. Finally, it may be stated that it is fundamental for neuroeducation to address the cordial virtues (Codina Felip 2015), thereby we need to draw on our knowledge of the neuroscientific research on humour, in order to develop a specific neuroeducation for ethical humour, one which would also include an education devoted to important ethical virtues.

6.9 Conclusions

Humour is the ability to perceive or show what is comical in the world around us, or what we sense within us, and it causes the emotion of mirth, which is expressed as a smile or laughter. Smiling and laughter are involuntary and arise deep within us, given that, in general, we cannot avoid laughing at what we find funny. However, smiling and laughter are voluntary in a profoundly ethical sense, given that they are reflections of the character we construct by exercising our own will and, as such, they can be trained. Since we can laugh in an ethically correct or incorrect manner, humour reflects an individual's ethical education.

"Ethical humour" is also therapeutic, in the sense that it is beneficial for one's health.

Its benefits include the following: stimulation of the immune system; stress reduction; increased endorphin secretion, which has analgesic effects; a reduced sense of distance between people; greater cohesion in workgroups; an ability to see the funny side of things; an improved feeling of well-being and increased self-esteem; greater capacity to express profound fears; distraction from threats and, specifically, distraction from the symptoms of pain. However, it is important to note that the humour that produces these positive effects is "ethical humour". If we do not employ this concept, our research on the effects of humour may result in contradictory findings. Indeed, some studies have indicated that the recourse to humour in coping conditions leads to emotional depletion and feelings of depersonalisation. However, these studies have been performed using scales that only measure the quantitative dimension of humour. Subsequent qualitative studies have shown that any negative effects are caused by "aggressive" humour, while humour when practised correctly does have positive effects.

Neuroscientific knowledge on humour has been gained by studying brain injuries, and through the use of electrocardiography and neuroimaging. These findings are essential for developing a neuroethics of humour. It has been shown that some lesions in the brain's right hemisphere cause personality changes, which lead to patients laughing inappropriately and offensively. The left hemisphere perceives incongruity and can therefore produce the comic, while the right hemisphere resolves any incongruity in a socially appropriate manner. The mesolimbic reward network also plays a role by developing the pleasant emotion of mirth.

To construct a "neuroethics of humour", the location for human freedom must first be established with regard to a key finding of neuroscientific studies, which is that on feeling the emotion of hilarity the muscles linked with laughter move involuntarily. However, as has been argued above this is not a justification for any lack of freedom related to laughter, because we can train our sense of humour so that the emotion of hilarity only arises when the comic object is truly ethical. We can also laugh voluntarily without experiencing the emotion of mirth, but this is fake laughter and, therefore, only of interest when thinking about etiquette (so as to not embarrass the individual who attempted to tell the joke). It is of little interest for an ethics of humour, which is based on the essence of our character.

The so-called "involuntary emotional smile" is that which activates the limbic system, while the "voluntary unemotional smile" activates areas of the frontal cortex. In regard to this, it should be highlighted that even when emotional laughter activates the limbic system (which does not occur with unemotional voluntary laughter), a number of studies concur that a large number of areas of the brain participate in authentic laughter, areas such as the cerebral cortex, the limbic system and the brainstem. The so-called "involuntary laughter" is, therefore, either a conceptual error, and it may be argued that there are many cases in which it is voluntary in a deep sense, or morally irrelevant. An example of morally irrelevant involuntary laughter is that caused by the electrical stimulation of an area of the left frontal lobe of the cortex, which produces the emotion of mirth without the corresponding cognitive component. I believe this is an example in which the personality is nullified in a manner similar to how drugs act, making the user laugh at, or become frightened of events that are not occurring. The ethical responsibility for such manipulations of a personality lie, in this case, with the researchers who conduct these techniques, just as it does with individuals who consume drugs that cause laughter. In any case, this type of laughter is not morally valuable because it is not a reflection of the deepest moral values of the individual who laughs.

We can stop laughing if we consider it inappropriate; for example, for immoral laughter. Neuroscientific studies have demonstrated that the cerebellum plays an important role in modulating the intensity and duration of laughter, and even inhibits it. However, the participation of the cerebellum does not necessarily exclude our ethical education from playing a role in the modification or inhibition of laughter. It is argued by some experts that listening to others laugh activates parts of the brain that in turn contribute to experiencing the emotion of mirth, by the so-called "mirror neuron" mechanism. However, in my view, while the laughter of others may contribute to our laughter it does not determine our laughter. We could see a group of people laughing at one another and if we considered it inappropriate, we could activate our cerebellum to inhibit our own laughter. The key lies in the ethical education of our humour.

To develop an ethics of humour, we should bear in mind that malicious laughter reflects the unethical character of the individual who laughs. Nonetheless, there is still a great deal to investigate in this field in order to clearly identify when laughter is malicious. It is still more important to clarify the entire positive dimension encompassed by ethical humour. One of the most relevant characteristics of ethical humour is that it is practised at the proper place and time and to the appropriate degree. I believe that ethical humour promotes a series of virtues that can also make a major contribution to improving health ethics. These virtues include patience, tolerance, kindness, humility, perseverance and courage. By locating humour within the more rational part of the brain, and mirth in the most emotional part, the importance of ethics education for humour become apparent. Humour avoids a purely emotional response, which, being thoughtless, could hurt others. Humour helps rationalise what has occurred and drawing on this rationalisation, goes on to form our ethical personality.

"Ethical" humour contributes to improving people's health, and we can develop this virtuous humour through education, and so become able to laugh at the appropriate time for the correct reasons to the appropriate degree. Thereby it is important that neuroscientific studies on humour employ the concept of "ethical humour", above all if they aspire to extend the application of their findings to the improvement of healthcare. In short, we need to develop a neuroeducation of ethical humour.

References

Aquinas, T. 1972. *Summa Theologiae*. Volume 44. *Well-Tempered Passion*. Trans. T. Gilby. New York: McGraw-Hill.

Aristotle. 1941. *The basic works of Aristotle*, ed. R. McKeon. New York: Random House.

Arroyo, S., R.P. Lesser, B. Gordon, S. Uematsu, J. Hart, P. Schwerdt, et al. 1993. Mirth, laughter and gelastic seizures. *Brain* 116: 757–780.

Azim, E., D. Mobbs, B. Jo, V. Menon, and A.L. Reiss. 2005. Sex differences in brain activation elicited by humor. *Proceedings of the National Academy of Sciences* 102 (45): 16496–16501.

Basu, S. 1999. 'Woe unto you that laugh now!': Humor and toleration in Overton and Shaftesbury. In *Religious toleration. "The varieties of rites" from Cyrus to Defoe*, ed. J.Ch. Laursen, 147–172. New York: St. Martin's.

Berk, L. et al. 1989. Neuroendocrine and stress hormone changes during mirthful laughter. *American Journal Medicines Science* 298: 390–396.
Bihrle, A.M., H.H. Brownell, and J.A. Powelson. 1986. Comprehension of humorous and nonhumorous materials by left and right brain-damaged patients. *Brain and Cognition* 5 (4): 399–411.
Bihrle, A.M., H.H. Brownell, and H. Gardner. 1988. Humor and the right hemisphere: A narrative perspective. In *Contemporary reviews in neuropsychology*, ed. H.A. Whitaker, 109–126. New York: Springer.
Black, D.W. 1984. Laughter. *Journal of American Medical Association. JAMA* 252: 2995–2998.
Blakemore, S.J., D.M. Wolpert, and C.D. Frith. 1998. Central cancellation of self- produced tickle sensation. *Nature Neuroscience* 1 (7): 635–640.
Bokun, B. 1989. *Humor theraphy: In cancer, psychosomatic diseases, mental disorders, crime, interpersonal and sexual relationships*. London: Riverrun.
Brownell, H.H., and H. Gardner. 1988. Neuropsychological insights into humour. In *Laughing matters: A serious look at humour*, ed. J. Durant and J. Miller, 17–34. Harlow: Longman Scientific and Technical.
Burns, T. 1953. Friends, enemies and polite fiction. *American Sociology Review* 28: 654–662.
Carbelo, B. 2008. *El humor en la relación con el paciente. Una guía para profesionales de la salud.* Barcelona: Elsevier–Masson.
Carver, C.S., M.F. Scheier, and J.K. Weintraub. 1989. Assessing coping strategies: A theoretically based approach. *Journal of Personality and Social Psychology* 56 (2): 267–283.
Champman, A.J., and H.C. Foot. 1996. *Humor and laughter. Theory, research and applications.* London/New York: Wiley.
Codina Felip, María José. 2015. *Neuroeducación en virtudes cordiales. Cómo reconciliar lo que decimos con lo que hacemos*. Barcelona: Octaedro.
Coser, R.L. 1959. Some social functions of laughter. *Human Relations* 12: 171–182.
Coulson, S., and M. Kutas. 2001. Getting it: Human event-related brain response to jokes in good and poor comprenhenders. *Neuroscience Letters* 316: 71–74.
Darwin, C. 1872. *The expression of the emotions in man and animals*. London: John Murray.
De Bono, E. 1993. *Serious creativity. Using the power of lateral thinking to create new ideas.* New York: HarperCollins.
De Sousa, R. 1987. When is it wrong to laugh? In *The philosophy of laughter and humor*, ed. John Morreall. Albany: State University of New York Press.
Derks, P., L.S. Gillikin, D.S. Bartolome-Rull, and E.H. Bogart. 1997. Laughter and electroencephalographic activity. *Humor: International Journal of Humor Research* 10 (3): 285–300.
Dixon, P.N., W.K. Willingham, C.K. Chandler, and K. McDougal. 1986. Relating social interest and dogmatism to happiness and sense of humor. *Individual Psychology Journal of Adlerian Theory, Research, and Practice* 42: 421–427.
Dorz, S., C. Novara, C. Sica, and E. Sanavio. 2003. Predicting burnout among HIV/AIDS and oncology health care workers. *Psychology and Health* 18 (5): 677–684.
Foley, E., R. Matheis, and C. Schaefer. 2002. Effect of forced laughter on mood. *Psychological Reports* 90 (1): 184.
Fox, R. 1953. *War F-second and the research physician*. Unpublished PhD dissertation, Harvard University.
Freud, S. 1963. Jokes and their relation to the unconscious. Standard Edition of the complete psychological works of Sigmund Freud, vol. 5. Trans. J. Strachey. New York: W.W. Norton.
———. 1987. Humor. Trans. J. Rivière. In *The Philosophy of Laughter and Humor*, ed. J. Morreall, 111–116. Albany: State University of New York Press.
Fried, I., C.L. Wilson, K.A. MacDonald, and E.J. Behnke. 1998. Electric current stimulates laughter. *Nature* 391 (66–68): 650.
Fry, W.F. 1977. The respiratory components of mirthful laughter. *Journal of Biological Psychology* 19: 39–50.
———. 1992. The psysiologic effects of humor, mirth, and laughter. *Journal of American Medical Association JAMA* 267: 1857–1858.

Gervais, M., and D.S. Wilson. 2005. The evolution and functions of laughter and humor: A synthetic approach. *Quarterly Review of Biology* 80 (4): 395–430.

Gillikin, L.S., and P.L. Derks. 1991. Humor appreciation and mood in stroke patients. *Cognitive Rehabilitation* 9 (5): 30–35.

Hobbes, T. 1839. Leviathan. In *The English works of Thomas Hobbes*, ed. W. Molesworth, vol. 3. London: Bohn.

Iwase, M., Y. Ouchi, H. Okada, C. Yokoyama, S. Nobezawa, E. Yoshikawa, et al. 2002. Neural substrates of human facial expression of pleasant emotion induced by comic films: A PET study. *NeuroImage* 17: 758–768.

Kant, I. 1892. *Critique of Judgment*. Trans. J.H. Bernard. London: Macmillan

Lipman, S. 1991. *Laughter in hell. The use of humor during the holocaust*. Northvale: Jason Aronson.

MacLean, P.D. 1987. The midline frontolimbic cortex and the evolution of crying and laughter. In *The frontal lobes revisited*, ed. E. Perecman, 121–140. New York: IRBN Press.

Martin, R.A. 2006. *The psychology of humor. An integrative approach*. Amsterdam: Elsevier.

Martin, R.A., P. Puhlik-Doris, G. Larsen, J. Gray, and K. Weir. 2003. Individual differences in uses of humor and their relation to psychological well-being: Development of the humor styles questionnaire. *Journal of Research in Personality* 37 (1): 48–75.

Matz, A., and S.T. Brown. 1998. Humor and pain management. A review of current literature. *Journal of Holistic Nursing* 16: 68–75.

McGhee, P.E. 1983. The role of arousal and hemispheric lateralization in humor. In *Handbook of humor research, vol. 1: Basic issues*, ed. P.E. McGhee and J.H. Goldstein, 13–37. New York: Springer.

Miller, G.R., and P. Bacon. 1971. Open-and-closed-mindedness and recognition of visual humor. *Journal of Communication* 21: 150–159.

Mobbs, D., M.D. Greicius, E. Abdel-Azim, V. Menon, and A.L. Reiss. 2003. Humor modulates the mesolimbic reward centers. *Neuron* 40: 1041–1048.

Mobbs, D., C.C. Hagan, E. Azim, V. Menon, and A.L. Reiss. 2005. Personality predicts activity in reward and emotional regions associated with humor. *Proceedings of the National Academy of Science* 102 (45): 16502–16506.

Morreall, J. 1987. *The philosophy of laughter and humor*. Albany: State University of New York Press.

———. 1997. *Humor works*. Amherst: Human Resource Development Press.

———. 2009. *Comic relief. A comprehensive philosophy of humor*. Malden: Wiley-Blackwell.

Parvizi, J., S.W. Anderson, C.O. Martin, H. Damasio, and A.R. Damasio. 2001. Pathological laughter and crying: A link to the cerebellum. *Brain* 124 (9): 1708–1719.

Plato. 1978. Philebus 48-50. In *The collected dialogues of Plato*, ed. E. Hamilton and H. Cairs. Princeton: Princeton University Press.

Porterfield, A.L., F.S. Mayer, K.G. Dougherty, K.E. Kredich, M.M. Kronberg, K.M. Marsee, and Y. Okazaki. 1988. Private self-consciousness, canned laughter and responses to humorous stimuli. *Journal of Research in Personality* 22: 409–423.

Roberts, R.C. 1988. Humor and the virtues. *Inquiry* 31: 127–149.

Sander, K., and H. Scheich. 2001. Auditory perception of laughing and crying activates human amygdale regardless of attentional state. *Cognitive Brain Research* 12 (2): 181–198.

Sayre, J. 2001. The use of aberrant medical humor by psychiatric unit staff. *Issues in Mental Health Nursing* 22 (7): 669–689.

Shammi, P., and D.T. Stuss. 1999. Humor appreciation: A role of the right frontal lobe. *Brain* 122 (4): 657–666.

Siurana, J.C. 2012. Filosofía del humor: estado de la cuestión. *Diálogo Filosófico* 82: 4–33.

———. 2015. *Ética del humor. Fundamentos y aplicaciones de una nueva teoría ética*. Madrid: Plaza y Valdés.

Stopsky, F. 1992. *Humor in the classroom. A new approach to critical thinking*. Lowell: Discovery.

Strickland, D. 1993. Seriously, laughter, matters. *Today's O.R. Nursing*, Nov.-Dic.: 19–24.

Svebak, S. 1982. The effect of mirthfulness upon amount of discordant right-left occipital EEG alpha. *Motivation and Emotion* 6 (2): 133–147.

Vaillant, G.E., and C.O. Vaillant. 1990. Natural history of male psychological health: XII. A 45-year study of predictors of successful aging at age 65. *American Journal of Psychiatry* 147: 31–37.

Weeks, A. 2000. Let's hear it for endorphins! *Online Newsletter*. www.healingpassages.com.

Wild, B., F.A. Rodden, W. Grodd, and W. Ruch. 2003. Neural correlates of laughter and humour. *Brain* 126 (10): 2121–2138.

Wooten, P. 1997. You've got to be kidding! Humor skills for surviving managed care. *Dermatology Nursing* 9: 423–429.

Yoshino, S., J. Fujimori, and M. Khoda. 1996. Effects of mirthful laughter on neuroendocrine and immune systems in patients with rheumatoid arthritis. *Rheumatology* 23: 793–794.

Yovetich, N.A., A. Dale, and M.A. Hudak. 1990. Benefits of humor in reduction of threat- induced anxiety. *Psychological Reports* 66: 51–58.

Chapter 7
The Uses of the Imagination in Moral Neur oeducation

Francisco Arenas-Dolz

7.1 Introduction

Views on moral education are today the subject of dispute. Two rival approaches have been developed, both rooted in philosophical paradigms: firstly, a universalist perspective with regard to justice, which addresses the correct course of action to take in a specific moral situation; and secondly, a particularist perspective concerned with virtue, and focused on agents and their cultivation of virtues.

Both Jean Piaget's theory of moral development (1932) and Lawrence Kohlberg's application of the Piagetian concept of the study of moral judgement (1981; 1984) favour reasoning as a means of defining logic and the thought schemas which govern moral understanding over other processes, such as affectivity, experienced by the individual in specific moral situations. The moral education methods derived from these approaches have advocated the development of universal structures of moral judgement, the acquisition of skills to enable a proper dialogue and make any judgement consistent with moral action, as well as a means to recognise and assimilate universal values. Furthermore, techniques have emerged, such as moral dilemmas, that favour the development of universal structures of moral judgement that in turn permit the adoption of general principles of value (Arenas-Dolz 2016a).

Besides the aforementioned authors, other studies have been undertaken on additional relevant aspects. John Bowlby (1988), drawing on his theory of attachment with its psychoanalytical roots, insisted on the emotional bond developed and established between two people, through reciprocal interaction, with the immediate aim of seeking and maintaining closeness. Carol Gilligan (1982) has highlighted the dimension of care. James Rest (1984) has studied the internal processes involved in the moral function and has proposed a model of ethical behaviour based on four components –the interpretation of a situation, calculation of what should be done,

F. Arenas-Dolz (✉)
Universidad de Valencia, Valencia, Spain
e-mail: Francisco.Arenas@uv.es

P. Calvo, J. Gracia-Calandín (eds.), *Moral Neuroeducation for a Democratic and Pluralistic Society*, https://doi.org/10.1007/978-3-030-22562-9_7

evaluation of the different courses of action and the execution of an action plan–, as well as an ability to process information, employ cognitive skills, respond to emotions and engage in social action. Richard Stanley Peters (1966, 1981) and Richard Knowles (1992) have proposed comprehensive models on moral development. Richard A. Shweder (1993), working in the discipline of cultural anthropology, has devised guidelines for studying the cultural variability of moral judgements and grouping them according to three paradigms: the *ethics of autonomy*, which considers the "self" as an autonomous being with rights; the *ethics of community*, which defines the "self" as a member of a social group, with duties and obligations; and the *ethics of divinity*, which is based on the "self" as a spiritual being. Dennis L. Krebs and Kathy Denton (2005) have proposed parallels between the evolution of cognitive tools and the first stages of Kohlberg's theoretical model; they consider post-conventional stages of moral education to be extra-evolutionary. Jonathan Haidt and Craig Joseph (2007) have drawn on social intuitionism to develop a *theory of moral foundations* which seeks to explain the origins and/or variations in moral reasoning, based on six innate, modular foundations: care/harm, justice/deception, freedom/oppression, loyalty/betrayal, authority/subversion and sanctity/degradation. Underpinned by moral judgements, through which a range of intuitive ethics would be developed and possessed by all individuals, these modular foundations are considered to have arisen over the course of human evolution as a response to the challenges of adaptation, and they have been further developed through culture. From the critical discussion of these perspectives on moral judgement, Darcia Narvaez (2008a, b) has formulated a theory of triune ethics, based on neuroscience, which proposes an approach to moral education based on motivational orientations that are rooted in evolved unconscious emotional systems, which have been formed by experience.

This contribution is a twofold one. Firstly, it offers a model of moral neuroeducation based on the theory of triune ethics, which links neurobiology, affective neuroscience and cognitive science, thereby demonstrating both its scope and limits. Secondly, it provides a detailed study of the role of the literary mind in the construction of knowledge and its potential as an interpretative instrument of and catalyst for environmental stimuli, which enables a deeper understanding of the world around us, as well as a means of structuring our experience, developing our capacity for evaluation and constructing various possibilities for action.

7.2 Triune Ethics and Moral Neuroeducation

Narvaez has summarised the results of her work in a *theory of triune ethics*. Inspired by Paul D. MacLean's *Theory of the triune brain* (1990), according to which the human brain has three interconnected systems, each with its own intelligence, subjectivity, sense of time and memory – the reptilian complex, the limbic system and the neocortex – Narvaez identifies three ethical orientations based on the evolution of these three structures. Associated with the reptilian complex, the *ethic of security* is focused on self-preservation and authority, which allows us to react to stressful or unexpected situations with a basic or instinctive response and so endeavour to ensure our survival. Associated with the limbic system, the *ethic of engagement* is focused on social bonds or emotional relationships with others; it requires a secure attachment founded on the stimulation and affection received in infancy. Finally, associated with the neocortex, found in the brains of more developed mammals, the *ethic of imagination* is focused on creative ways of thinking and acting socially; it allows us to consider alternative courses of action when making moral judgements and provides us with the capacity for problem solving, memorisation and deliberative learning. The three cerebral neurobiological systems underlying the three ethics illustrate the aspects of development that lead to an optimal mode of functioning: a sense of security, a sense of active or committed participation to a social life, and a sense of the imagination's role in attaining positive ends. A morally *exemplary* person would be capable of coordinating these three systems (Narvaez 2008a, b, 2014, 2016).

The capacity for deliberating on morality allows us to reflect on what courses of action to follow, and education provides us with the means for this deliberative learning (Arenas-Dolz 2008, 2009a, b, 2013a). However, the *ethic of imagination* also operates through tacit knowledge, learned through experience over the course of one's life. Thus, moral judgement involves the coordination of the *deliberative or conscious mind* and the *intuitive or unconscious mind*. The *ethic of imagination* provides a greater moral sense than the other ethics, yet, if the *ethic of engagement* and *ethic of imagination* have not been sufficiently nurtured by the family or community, it is the *ethic of security* that operates by default. On this basis, Narvaez distinguishes three types of moral imagination: *detached imagination*, focused on the search for procedures to resolve abstract moral problems, regardless of the context; *vicious imagination*, focused on the ego and motivated by a search for power; and *communal imagination*, capable of developing a sense of connection that extends beyond any immediate relationships, and which uses the skills of the mind in a prosocial way. It is the latter that is the basis for a fully developed morality, one that includes the use of abstract capacities, adequately based on experience, in order to resolve moral problems, such as relationships with others. A morality developed in this way can neutralise emotional reactions generated in the oldest parts of the brain (Narvaez and Mkrva 2014).

Inspired by Rest's four-component model and equidistant between both rationalism and intuitionism, due to its lack of adaptation to reality, Darcia Narvaez (2006) has developed a model of integrative ethical education that bridges the aforementioned universalist and intuitionist theoretical outlooks. Her educational model aims to teach students how to acquire moral skills and it is based on an accompanied, or monitored practice, undertaken at four levels: immersion in examples and opportunities (addressing the general context, learning to recognise basic patterns); attention to facts and skills (considering details and prototypical knowledge; accumulating knowledge); practising procedures (establishing goals, planning steps for problem solving, practising skills); and integrating knowledge and procedures (executing plans, problem solving).

Narvaez has identified 84 different skills involved in moral action. Experts' knowledge of these skills, according to the writer, is tacit knowledge, which allows them to make decisions quickly, even automatically. The learning processes allow students to become more skilful in the use and adaptation of knowledge in order to respond to increasingly complex problems. The teacher must incentivise students to put themselves in the place of others so as to perceive new contexts and points of view based on their own experiences, thereby stimulating an interaction between the students' tacit knowledge and explicit knowledge. The outcome of Narvaez's research is three ideas that have significant educational implications: moral development is best interpreted as the development of moral skills; interactive education is transformative by nature; and human nature is cooperative and self-actualised. According to this holistic and Aristotelian model, although the author does not explicitly define it as the latter, the importance of the community for attaining virtue and human happiness is vindicated, and experts possess the practical wisdom needed to act in each situation (Narvaez 2013).

In subsequent contributions, Narvaez (2008b: 290) has initiated a critical dialogue with neuroscience from the perspective of evolutionary psychology. Although neuroscience may be able to contribute to moral psychology in three ways—a healthy moral development requires the appropriate cerebral functions; studies on cerebral function could corroborate or question traditional theories; and surgical interventions on the brain may be able to contribute to moral improvement—the writer distances herself from the latter opinion, as the current state of our knowledge on cerebral function is still in its early stages. Nonetheless, she incorporates two contributions from neuroscience in her work. Firstly, drawing on neuroscience, she identifies the parts of the brain related to the four components of her educational model (Narvaez 2008b: 296–300). Secondly, she focuses on cerebral plasticity and, taking into account the growth of neurons over the course of a lifetime, considers that the conditions of childhood may prove crucial for subsequent development (Narvaez 2008b: 300–303; Narvaez et al. 2013, 2014).

7.3 The Social Nature of Moral Imagination

A further pertinent dimension of Narvaez's extensive is her consideration of the *social nature* of moral imagination. John Dewey indicated that the imagination plays a central role in deliberation by mentally representing different possibilities and alternatives. For Dewey, deliberation is "a dramatic rehearsal (in imagination) of various competing possible lines of action [...] Deliberation is an experiment in finding out what the various lines of possible action are really like. It is an experiment in making various combinations of selected elements of habits and impulses, to see what the resultant action would be like if it were entered upon. But the trial is in imagination, not in overt fact. The experiment is carried on by tentative rehearsals in thought which do not affect physical facts outside the body. Thought runs ahead and foresees outcomes, and thereby avoids having to await, the instruction of actual failure and disaster. An act overtly tried out is irrevocable, its consequences cannot be blotted out. An act tried out in imagination is not final or fatal. It is retrievable" (Dewey 1922: 190).

As the quality of any deliberation depends on an individual's capacity for imagining alternative, different or novel lines of action, the cultivation of the imagination is crucial for the development of intelligence. Dewey writes, "Were all instructors to realize that the quality of mental process, not the production of correct answers, is the measure of educative growth something hardly less than a revolution in teaching would be worked" (Dewey 1916: 207). The revolution will take place when teachers realise that coming up with the correct answer is not a measurement of educational growth, instead it is the quality of mental processes that should be attended to. Dewey sought a middle ground between the two influential movements of his era. The first movement understood school as the place where students mature, focusing on the curriculum and a range of knowledge that students had to learn in a very passive manner. For education to be effective, such knowledge must be aligned with the students' interests. The second movement downplayed both the role of the teacher and the importance of learning and focused solely on the child, which was also counter-productive, according to Dewey. For him, it was necessary to strike a balance, with the curriculum on one hand, and the students on the other, as the two extremes of the same process. Combining both would link experience with thought instead of solely being concerned with using the brain. It would be only in this way that the senses could recover their value: by putting them in direct contact with reality, the abstract is converted into something specific and requires our greater involvement, which in turn makes us more active and involved in specific tasks. Dewey's ideas on education have provided a foundation for problem-based learning and are often used today in facilitating the acquisition of learning an active pursuit, as opposed to the passive absorption of facts. With traditional forms of learning, the teacher is in charge of the identification of learning needs and the explanation of knowledge. In contrast, with teaching based on problem or challenge-solving, or else project-work, the teaching method involves the teacher guiding a group of students so that they actively, constructively, creatively and autonomously find the answer to a question or the solution to a problem; they have to seek, understand, integrate, and apply the basic concepts of the problem, as well as related concepts, in order to correctly solve the complete the task at hand. This active methodology combines the capacity for reasoning and autonomy of judgement and is thereby fundamental for teaching philosophy. In line with Dewey, Martha Nussbaum (2010) has insisted on the role played by the imagination and the teaching of the humanities in enhancing the quality of democracy. According to Nussbaum, an imaginative intelligence is required for deliberation in order to achieve a democratic purpose, and this form of intelligence is the outcome of both a habit and a skill that must be cultivated (Gil Blasco 2014; Modzelewski Drobniewski 2012; Tienda Palop 2011, 2015).

7.4 Neurobiology, Affective Neuroscience and Cognitive Science

Over the course of her research, Narvaez has also addressed recent discoveries made in cognitive science. Firstly, she has considered the importance of the affective roots of culture and cognition set out in Antonio Damasio's *theory of somatic markers* (1999), which is a theory of the mind focused on the body that seeks to explain the influence of the emotions on decision-making and reasoning processes. According to Damasio (1994) emotions are represented in the theatre of the body, while feelings are represented in the theatre of the mind; emotions are physical and feelings are mental; emotions can be experienced by different species, but feelings seem to be reserved for the human species, which has a mind – animals experience emotions, but do not feel them. This theory has become an essential element in many studies on language, especially with regard to the *theory of the conceptual metaphor* (Lakoff and Johnson 1980).

Secondly, Narvaez has drawn on Mark Johnson's *theory of the embodied imagination* (1987, 1993), as a basis for a critique of ethical theories that present morality as a universal system of laws or rules based on reason. On the basis of his observation that "moral reasoning is a constructive imaginative activity" (Johnson 1993: 2), Johnson argues for the need to change our concept of moral reasoning, stating that it is not right to think and act as if we possessed a universal, disembodied reason that generates absolute rules, procedures for decision-making, and laws or universal categories that enable us to distinguish good from evil in any situation we may find ourselves in. He writes, "An exclusive focus on moral laws and rational principles is a threat to human well-being because it blinds us to the cultivation of moral imagination that is necessary if we are to be morally sensitive and fully responsible to other people. The narrow and simplistic conception of morality as the following of moral laws completely ignores the crucial imaginative dimensions of moral understanding that make it possible for us to discern what is important in any situation or relationship and to act wisely in light of our discernment" (Johnson 1993: 5). According to Johnson, "[paying] attention to the imaginative dimensions of our understanding will give us a *new* set of questions to investigate that couldn't even be asked under the influence of our previous assumptions" (Johnson 1993: 12).

Thirdly, Narvaez's work has engaged with the *theory of conceptual integration*, proposed by Gilles Fauconnier and Mark Turner (2002), as a cognitive tool for understanding how human thought operates within conceptual networks and carries out a variety of discursive functions, which range from the integration of cases, conceptual change and metaphorical projection to humour, literary invention and the transmission of feelings and attitudes. According to this theory, the participants of a discourse construct mental spaces for the purpose of comprehension and local action. As Mark Turner (1996, 2001) has demonstrated, the description of intellectual figures or tropes offered to us by classical rhetoric is not solely restricted to literary and poetic texts, but also encompasses the common and natural way in which all human beings speak and form knowledge; this form of description may be far better adapted to the analysis of cognitive aspects of the human nervous system, than any other description of human language. The presence of rhetoric in the field of neuroscience is not at all outlandish (Arenas-Dolz 2012, 2013b, 2016b). The use of rhetoric has proved to be significant in the field of neuroscience, especially in the school of American neurobiologist Gerald Edelman (Edelman 1992; Edelman and Tononi 2000). The binary system used by neo-Saussurian linguistics to describe language is insufficient when considering human thought and language. The human brain is not a computer. For Turner, a scholar of the neurobiology of cognition, the "literary mind" – the forms of representations of stories and parables and, ultimately, any other narrative form – is the basis for our ways of thinking, the central principle of our experience and moral knowledge, and an indispensable tool of reason that allows us to construct mental spaces in which we can understand and act. We use narrative forms to grasp what is located in space and in time, to conceive of ourselves and understand the ways others think.

Conceptual integration is an important source of creativity in human beings and it enables the analysis of patterns that occur during the creative process, which are based on selectively combining pre-existing ideas. Creative fusion, with its new, emerging structure, allows us to understand the truth. Although it is commonly thought that the "fictitious" is the opposite of the "real", for Turner, fiction is indispensable for the human condition. The creative fusion may be fictitious, but it is offered as a tool for understanding the truth (Turner 2006, 2014). Through the description of the cognitive processes of common language, Turner has overcome the antithesis between "figurative meaning" and "literal meaning", between "truth" and "degrees of verisimilitude".

7.5 The Cultivation of Moral Imagination

In a study on the current trends and approaches of media literacy in Europe, the European Commission (2007: 13) defined *media literacy* as "the ability to access, analyse and evaluate the power of images, sounds and messages which we are now being confronted with on a daily basis and which are an important part of our contemporary culture; as well as to communicate competently using media available, on a personal basis. Media literacy relates to all media, including television and film, radio and recorded music, print media, the internet and other new digital communication technologies". The cultivation of imagination is today more necessary than ever.

The use of literary works for contextualising situations and cultivating the skills of imagination is essential for the evolution of the qualitative distinctions among human beings. The power of imagination as a human capacity allows us to create new perspectives of the world. Human solidarity is not achieved through academic research, but through the imaginative capacity of seeing others as companions undergoing suffering. An important function of literature is to challenge our dominant self-descriptions, thereby expanding our individual and collective sense of self. The process of seeing other human beings as "one of us" instead of "them" is not a matter to be addressed through theory, but through the diverse genres that grant the novel a special place as a vehicle for change and moral progress. *Media literacy* is an important component for constructing and maintaining democracy.

In opposition to Aristotle, Kant in his *Critique of judgement*, both in the Prologue and in § 60, formulates the thesis that aesthetic judgements do lead to the discovery of things. For Kant, morality is fundamentally based on reason, not on imagination. Freely exercised, the imagination is a more independent faculty, which reveals good and bad ways of imagining, as well as creative and mechanical, conscious and unconscious, ordinary and superior ways. Thus, unlike Hegel, for whom taste, although lacking a principal aim, allows things to be represented as if they were real, Kant in *The critique of pure reason*, considers imagination as an art, hidden in the depths of the human soul, whose true modes of action can only be discovered and unveiled with difficulty (*KrV* A 141/B 180, Kant 1971). The separation of moral questions from questions of taste has since prevailed, which has led to the discussion of the cognitive value of literature, its practical dimension and its role in the context of philosophy (Camps 1983; López de la Vieja 2003).

Moral imagination requires: the capacity to imagine ourselves in a network of relationships which includes our enemies; the ability to nourish a contradictory curiosity that encompasses complexity; a firm belief in, and quest for the creative act; the acceptance of the risk of advancing towards the unknown – ultimately, courage.

Courage is imaginative and arises from the ability to act. It is to the good that we have a moral imagination, provided we do not treat it as an absolute. Not all moral images—the way in which our ideas and virtues are related—are equally good. There are good and bad ways of imagining. However, we can learn from moral images on the basis of which certain people have led good lives, and they can likewise learn from ours. Through the use of the moral imagination, the impossible becomes possible in the form of an event, an unplanned, unpredictable action, as a response and a responsibility. The American literary critic Lionel Trilling summarised the potential of the moral imagination regarding the case of the novel, as follows:

"For our time the most effective agent of the moral imagination has been the novel of the last 200 years. It was never, either aesthetically or morally, a perfect form and its faults and failures can be quickly enumerated. But its greatness and its practical usefulness lay in its unremitting work of involving the reader himself in the moral life, inviting him to put his own motives under examination, suggesting that reality is not as he sees it. It taught us, as no other genre ever did, the extent of human variety and the value of variety. It was the literary form to which the emotions of understanding and forgiveness were indigenous, as if by the definition of the form itself. At the moment its impulse does not seem strong, for there never was a time when the virtues of its greatness were so likely to be thought of as weaknesses. Yet there never was a time when its particular activity was so much needed, was of so much practical, political and social use– so much so that if its impulse does not respond to the need, we can be sad not only over a waning form of art but also over a waning freedom" (Trilling 1948: 27).

For Simone Weil, imagination is the faculty opposed to *attention*. If attention allows the object to be fully captured and an individual to discriminate between the real and the deceptive (Weil 2002: 116–122), the imagination can hinder contact with reality, leading individuals to be deceived and entertain dangerous utopias. Therefore, "we must continually suspend the work of our imagination filling the void within ourselves. If we accept no matter what void, what stroke of fate can prevent us from loving the universe? We have the assurance that, come what may, the universe is full" (Weil 2002: 18). However, in Weil's later writings, the imagination is presented as a vital challenge, focusing on its involvement in insight, reading and attention (Weil 2004), and this signals what was a recurring issue for Hannah Arendt: the need for the imagination within history; the imagination permits the reconstruction of a history that has only been told by the victors; it takes into account the possibility of examining this history from other perspectives; and finally, it enables reflection on the distinctive nature of engaging in a dialogue within a public space. The opening up of the imagination evokes scenarios and projects prospects through which thinking about moral action becomes easier. Arendt distinguishes between *imagination* and *empathy*: the latter involves an identification with the feelings of another and is often equivalent to assuming their prejudices; in contrast, the imagination seeks to understand the other without relinquishing the distance required to make judgements for oneself (Arendt 1992: 43, 79–85). For Arendt, who thinks of imagination as intrinsically political, it is necessary to re-evaluate the moral imagination with specific regard to the forms of community life: "Imagination alone enables us to see things in their proper perspective, to be strong enough to put that which is too close at a certain distance so that we can see and understand it without bias and prejudice, to be generous enough to bridge abysses of remoteness until we can see and understand everything that is too far away from us as though it were our own affair" (Arendt 1994: 323).

Imagination as the source of understanding and action is an idea that was also explored by Iris Murdoch, for whose thought the imagination was a central concept. Literature may be an excellent means for stimulating and nourishing moral imagination. It is also a reflection of moral life and is capable of ridding us of the distorted images of both reality and ourselves, the products of our fantasy. In opposition to fantasy is the imagination, the creative imagination, which is a multiple capacity that seeks the truth in a distinctive manner to both philosophy and science (Murdoch 1961, 1982). This new way of seeing moral problems focuses on the way that language, metaphors and imagination affect our beliefs. It involves the use of narrative and life stories, rather than concepts and theory, to organise our experience and find meaning for it. In the article "Ethics and the imagination", originally published in 1986 in issue 52 of the magazine *The Irish Theological Quarterly*, Murdoch writes: "The work of imagination in art may be seen as a symbol of its operation elsewhere; this might also be expressed by saying that there is artistry in the sorting, separating and connecting movement of the mind in other areas, in science and scholarship, and in morals and politics where an ordering activity is fused with an ability to picture what is quite other; specially of course to picture and realise, make real to oneself, the existence and being of other people. Imagination in politics: to imagine the consequences of policies, to picture what it is like for people to be in certain situations (unemployed, persecuted, very poor), to relate axiomatic moral ideas (for instance about rights) to pragmatic and utilitarian considerations" (Murdoch 1992: 321–322).

Thus, seeing others, *all* others, as human beings, is an ethical task that each one of us is called on to assume. The capacity of seeing others as human beings depends on our capacity for paying attention to details, the intricacies, and putting ourselves in others' shoes through the imaginative activity, as has been demonstrated by Cora Diamond. Such a use of the imagination is a fundamental ethical task, for which it is necessary to cultivate a "creative sensibility", or in other words, to transform our perception of things and others while countering the education received and the cultural conventions of our social universe (Diamond 1991: 313).

The cultivation of the moral imagination is possible not only through literary examples, but also cinematic ones. To conclude, we may examine two emblematic works of cinema: *Dekalog* (1988–89) and the *Three Colours* trilogy (1993–94) by the Polish director and scriptwriter Krzysztof Kieślowski. In the series *Dekalog*, each of the ten episodes, approximately 55 min long, tells a story loosely based on one of the biblical ten commandments. The meaning of each commandment is explored through a fictional storyline that demonstrates the complexity and paradoxical nature of its relationship with each law through situations removed from reality. In the *Three Colours* trilogy Kieślowski takes inspiration from the colours of the French flag and the three revolutionary ideas they represent: blue (liberty), white (equality) and red (fraternity). The three films are meant to be seen separately, but when viewed as a whole they reveal numerous connections with each other, or rather, become elements of a work designed as a whole in which each utopic theme must be read in relation with the others. Illustrative of this sense is a recurring scene in the three films: an old woman faced by the difficulty of correctly disposing of her rubbish in colour-coded refuse bins. In the first film (*Blue*), Julie does not even realise that the old woman needs help. She has her eyes closed and her face towards the sun; she is young, her future lies ahead of her and she cannot even imagine old age and its problems. In the second film (*White*), Karol sees old age, recognises the difficulty it brings (he is also beset by many problems), but this does not result in any act of solidarity. Only in the third film (*Red*), which is dedicated to fraternity for good reason, does Valentine stop to help the old woman.

The concept of imagination is an essential concept because it can clarify the sense in which "we are all artists". However, and without detracting the value of the moral imagination, it is necessary to question the origin of the human need to have moral rules. It was Nietzsche in aphorism 188 of *Beyond good and evil* who considered that the long-term subjection to rules– in all senses, not just morality – was "the disciplinary means whereby the European spirit has attained its strength, its remorseless curiosity and subtle mobility" (Nietzsche 2017: 88). In other words, there is something in our moral and cognitive structure that fosters the image of rules and makes it viable, and before trying to replace it with another based solely on the imagination, the consequences of such a change must be considered with great care. Ultimately, a theory of imagination requires a theory of conscience. We do what we are. We can change what we are, but we cannot do so quickly or easily. Living involves following guidance, patterns, rules, finding the place of imagination and metaphor in our lives (Geary 2012).

7.6 Conclusions

Over the course of this chapter I have explored the possibility of overcoming overly narrow concepts of human thought and language, which reduce the moral imagination to a mere neurobiological phenomenon, by instead developing a focus upon the human capacity to generate, mobilise, construct and cultivate the moral imagination in order to create scenarios of both a desired and an undesired future. The possibility of overcoming a concept of education reduced to the exclusive scope of neuroscientific thought and human language is forged through our capacity for generating, mobilising and constructing a moral imagination. When the art of imagination is used to structure the world of moral judgements, it becomes the art of *evaluation*, the art of *attention*, through which we act and act anew in the world according to what moves our desires and reflections.

Triune ethics theory proposes an analysis of moral judgement which culminates in an *ethic of imagination* oriented towards improving moral development. This theoretical framework helps us to understand the principles of the mechanisms of learning and the basic components of human action; it allows us to understand the complex cognitive skills taught through education; it offers us an empirical foundation upon which to corroborate the knowledge present in teaching and to refute other knowledge; it contributes to a more precise understanding of learning mechanisms and provides us with a unified framework for understanding moral neuroeducation. However, a comprehensive model of moral education cannot be inscribed within a theory that subscribes to a scientific foundation in neurobiology. Drawing on the study of the moral imagination, it is necessary to take a step back from the theories that Descartes saw as a threat to the imagination and also those theories that reduce imagination to biology, which implies an unacceptable naturalisation of human sociability, and seeks to exempt the individual from the responsibility of making choices and is also incapable of avoiding a drift towards indifference. Even if the creative education of students could improve their moral problem solving skills, it may be argued that it is insufficient to use stimulation techniques based on neuroscientific studies on moral imagination. Focusing solely on such techniques may in fact isolate students from having to face the complexity of the everyday (Arenas-Dolz 2016a). Instead, a model of complementarity is required, one which recovers an open, multidisciplinary perspective for the analysis of issues of a certain complexity. What is required is the creation of a "research community" in which students can play an active and reflective role and apply new ideas.

To sum up, by conceiving the moral imagination in terms of triune ethics theory, the important heuristic role played by the imagination is underscored, and this gives rise to a process of critical reflection that mobilises arguments questioning our vision of certain events, and also shows how real experiences have not been considered in full. The complexity of practical issues requires more complex procedures in order to fully understand the variety of experiences and ways of life that characterise pluralistic societies. Triune ethics theory, based on the theory of the triune brain, is of key importance for its simplicity and clarity. However, it assumes, firstly, an abrupt separation between emotion and rationality, which are located in different parts of the brain despite both being intimately interrelated in our cerebral organisation and our mental operation; and secondly, it implies that the emergence of new and complex brain structures is the result of a process of growth due to a material expansion, although we now know that this is due to a reorganisation of existing circuits, and through which they have taken on greater complexity and new functions (Smith 2010).

In conclusion, a reconsideration of the relationships between ethics and imagination today means going beyond theories that reduce the moral imagination to its neurobiological bases and instead considering other perspectives on the moral imagination, ones which explore how art and literature can help improve our capacity for moral judgement. Practising the act of imagining the other (and implicitly oneself) is a necessary, impossible and endless task which leads human beings to "connect with the feelings and interests of others", to "feel and know they are obliged to mutually support each other in order to advance projects for a life worth living" (Cortina 2007: 206, 215).

References

Arenas-Dolz, Francisco. 2008. Retórica aristotélica y democracia deliberativa. *Sistema. Revista de Ciencias Sociales* 206: 91–110.

———. 2009a. Deliberatio siue consilium. *Anuari de Filologia. Secció D. Studia Graeca et Latina* 12: 9–33.

———. 2009b. Il luogo delle passioni nella deliberazione. Ciò che la democrazia deliberativa può apprendere dalla retorica aristotelica. *Filosofia Politica* 23 (3): 453–573.

———. 2012. Neuro-retórica. ¿Vino viejo en odres nuevos? In *Guía Comares de Neurofilosofía Práctica*, ed. Adela Cortina, 125–145. Granada: Comares.

———. 2013a. Deliberative democracy, active citizenship and critical culture. From Aristotle's *Rhetoric* to contemporary political philosophy. *Icono14.* Revista Científica de Comunicación y Tecnologías Emergentes 11(1): 163–193.

———. 2013b. ¿Qué es la neurorretórica? *Daímon.* Revista Internacional de Filosofía 58: 69–80.

———. 2016a. Neurofilosofía, naturalismo y juicio moral. In *De simios, cyborgs y dioses. La naturalización del hombre a debate*, ed. Claudia Carbonell and Lourdes Flamarique, 207–223. Madrid: Biblioteca Nueva.

———. 2016b. Cognición y retórica: bases biológicas del significado y la comprensión. *Pensamiento. Revista de Investigación e Información Filosófica* 273: 997–1018.

Arendt, Hannah. 1992. In *Lectures on Kant's political philosophy*, ed. Ronald Beiner. Chicago: The University of Chicago Press.

———. 1994. Understanding and politics. In *Essays in understanding, 1930–1954. Formation, exile, and totalitarianism*, ed. Jerome Kohn, 307–327. New York: Schocken Books.

Bowlby, John. 1988. *A secure base. Parent-child attachment and healthy human development*. London: Routledge.

Camps, Victoria. 1983. *La imaginación ética*. Barcelona: Seix Barral.

Cortina, Adela. 2007. *Ética de la razón cordial. Educar en la ciudadanía en el siglo XXI*. Oviedo: Ediciones Nobel.

Damasio, Antonio. 1994. *Descartes' error. Emotion, reason and the human brain*. New York: G.P. Putnam.

———. 1999. *The feeling of what happens. Body and emotion in the making of consciousness*. San Diego: Harcourt.

Dewey, John. 1916. *Democracy and education. An introduction to the philosophy of education*. New York: Macmillan.

———. 1922. *Human nature and conduct. An introduction to social psychology*. New York: H. Holt.

Diamond, Cora. 1991. *The realistic spirit. Wittgenstein, philosophy, and the mind*. Cambridge, MA: MIT Press.

Edelman, Gerald M. 1992. *Bright air, brilliant fire. On the matter of the mind*. New York: Basic Books.

Edelman, Gerald M., and Giulio S. Tononi. 2000. *A universe of consciousness. How matter becomes imagination*. New York: Basic Books.

European Commission. 2007. *Study on the current trends and approaches to media literacy in Europe*. Brussels: European Commission.

Fauconnier, Gilles, and Mark Turner. 2002. *The way we think. Conceptual blending and the mind's hidden complexities*. New York: Basic Books.

Geary, James. 2012. *I is an other. The secret life of metaphor and how it shapes the way we see the world*. New York: Harper Perennial.

Gil Blasco, Marta. 2014. *La teoría de las emociones de Martha Nussbaum. El papel de las emociones en la vida pública*. Ph.D. dissertation. Valencia: Universitat de València.

Gilligan, Carol. 1982. *In a different voice. Psychological theory and women's development*. Cambridge, MA: Harvard University Press.

Haidt, Jonathan, and Craig Joseph. 2007. The moral mind: How five sets of innate intuitions guide the development of many culture-specific virtues, and perhaps even modules. In *The innate mind, Vol. 3: Foundations and the future*, ed. Peter Carruthers, Stephen Laurence, and Stephen Stich, 367–391. New York: Oxford University Press.

Johnson, Mark. 1987. *The body in the mind. The bodily basis of meaning, imagination and reason*. Chicago: The University of Chicago Press.

———. 1993. *Moral imagination. Implications of cognitive science for ethics*. Chicago: The University of Chicago Press.

Kant, Immanuel. 1971. *The critique of pure reason. The critique of practical reason and other ethical treatises. The critique of judgement*. Chicago: Encyclopaedia Britannica.

Knowles, Richard. 1992. The acting person as moral agent: Erikson as the starting point for an integrated theory of moral development. In *Foundations of moral education: Vol. 2. Psychological foundations of moral education and character development. An integrated theory of moral development*, ed. George F. McLean and Richard Knowles, 219–250. Washington, DC: The Council for Research in Values and Philosophy.

Kohlberg, Lawrence. 1981. *The philosophy of moral development. Moral stages and the idea of justice*. San Francisco: Harper and Row.

———. 1984. *The psychology of moral development. The nature and validity of moral stages*. San Francisco: Harper and Row.

Krebs, Dennis L., and Kathy Denton. 2005. Toward a more pragmatic approach to morality: A critical evaluation of Kohlberg's model. *Psychological Review* 112 (3): 629–649.

Lakoff, George, and Mark Johnson. 1980. *Metaphors we live by*. Chicago: The Chicago University Press.

López de la Vieja, María Teresa. 2003. *Ética y literatura*. Madrid: Tecnos.

MacLean Paul, D. 1990. *The triune brain in evolution. Role in paleocerebral functions*. New York: Plenum.

Modzelewski Drobniewski, Helena. 2012. *La educabilidad de las emociones y su importancia para el desarrollo de un ethos democrático. La teoría de las emociones de Martha Nussbaum y su expansión a través del concepto de autorreflexión*. Ph.D. dissertation. Valencia: Universitat de València.

Murdoch, Iris. 1961. Against dryness. A polemical sketch. *Encounters* 16 (1): 16–20.

———. 1982. Philosophy and literature. In *Men of ideas. Some creators of contemporary philosophy*, ed. Bryan Magee, 262–284. Oxford: Oxford University Press.

———. 1992. *Metaphysics as a guide to morals*. London: Chatto & Windus.

Narvaez, Darcia. 2006. Integrative ethical education. In *Handbook of moral development*, ed. Melanie Killen and Judith G. Smetana, 703–733. Mahwah: Erlbaum.

———. 2008a. Triune ethics. The neurobiological roots of our multiple moralities. *New Ideas in Psychology* 26 (1): 95–119.

———. 2008b. Moral development and behaviour under the spotlight of the neurobiological sciences. *Journal of Moral Education* 37 (3): 289–312.

———. 2013. Wisdom as mature moral functioning: Insights from developmental psychology and neurobiology. In *Toward human flourishing. Character, practical wisdom and professional formation across the disciplines*, ed. Mark Jones, Paul Lewis, and Kelly Reffitt, 24–40. Macon: Mercer University Press.

———. 2014. Neurobiology and the development of human morality. In *Evolution, culture and wisdom*. New York: W.W. Norton.

———. 2016. *Embodied morality. Protectionism, engagement and imagination*. New York: Palgrave Macmillan.

Narvaez, Darcia, and Kellen Mkrva. 2014. Creative moral imagination. In *The ethics of creativity*, ed. Seana Moran, David H. Cropley, and James C. Kaufman, 25–45. New York: Palgrave Macmillan.

Narvaez, Darcia, Jaak Panksepp, Allan N. Schore, and Tracy R. Gleason, eds. 2013. *Evolution, early experience and human development. From research to practice and policy*. Oxford: Oxford University Press.

Narvaez, Darcia, Kristin Valentino, Agustin Fuentes, James J. McKenna, and Peter Gray, eds. 2014. *Ancestral landscapes in human evolution. Culture, childrearing and social wellbeing*. Oxford: Oxford University Press.

Nietzsche, Friedrich. 2017. *The essential Nietzsche. Beyond good and evil. The genealogy of morals*. Trans. Helen Zimmern. New York: Chartwell Books.

Nussbaum, Martha C. 2010. *Not for profit. Why democracy needs the humanities*. Princeton: Princeton University Press.

Peters, Richard Stanley. 1966. *Ethics and education*. London: George Allen and Unwin.

———. 1981. *Moral development and moral education*. London: George Allen and Unwin.

Piaget, Jean. 1932. *The moral judgment of the child*. London: Kegan Paul, Trench, Trübner and Co.

Rest, James. 1984. The major components of morality. In *Morality, moral behavior and moral development*, ed. William M. Kurtines and Jacob L. Gewirtz, 24–38. Toronto: Wiley-Interscience.

Shweder, Richard A. 1993. *Thinking through cultures. Expeditions in cultural psychology*. Cambridge, MA: Harvard University Press.

Smith, Chris U.M. 2010. The triune brain in antiquity: Plato, Aristotle, Erasistratus. *Journal of the History of the Neurosciences* 19 (1): 1–14.

Tienda Palop, Lidia. 2011. *El modelo de racionalidad de Martha C. Nussbaum. Emociones, capacidades y justicia*. Ph.D. dissertation. Valencia: Universitat de València.

———. 2015. El papel de las emociones y la literatura en la deliberación pública. La figura del equilibrio perceptivo de Martha C. Nussbaum. *Arbor* 191 (773): a241.

Trilling, Lionel. 1948. Manners, morals, and the novel. *The Kenyon Review* 10 (1): 11–27.
Turner, Mark. 1996. *The literary mind. The origins of thought and language*. Oxford: Oxford University Press.
———. 2001. *Cognitive dimensions of social science*. Oxford: Oxford University Press.
———. 2006. *The artful mind. Cognitive science and the riddle of human creativity*. Oxford: Oxford University Press.
———. 2014. *The origin of ideas. Blending, creativity, and the human spark*. Oxford: Oxford University Press.
Weil, Simone. 2002. *Gravity and grace*, trans. Emma Crawford and Mario von der Ruhr. London: Routledge.
———. 2004. *The notebooks of Simone Weil*, trans. Arthur Wills. London: Routledge & Kegan Paul.

Chapter 8
Overcoming Neuroessentialism. Towards an Integral Notion of Subjectivity for Moral Neuroeducation

Marina García-Granero

8.1 Introduction

Moral neuroeducation applies the insight and knowledge provided by neuroscience in order to effectively address the moral demands of our time. In so doing, this new field offers a new sense of possibility to the task of wisely educating the citizens of the twenty-first century on how to respect human rights, participate in political life and take responsibility for choices and actions. Codina has eloquently stated that the aim of moral neuroeducation is to reconcile what we say with what we actually do, or in other words, to act in accordance with our moral intentions, and the values that we publicly defend (Codina 2015).

Having said that, it is my opinion that moral neuroeducation should also tackle the impact of neuroscience on our self-understanding as human beings, and how it is changing our idea of personhood, which has far-reaching implications upon our world-view. For some years now, neuroculture has been the subject of discussion, as a culture resulting from the increasing assimilation of neuroscientific concepts into the cultural imaginary (Frazzeto and Anker 2009). These concepts are portrayed in literature, film, art, mass media, commercial products, and further afield across a culture that is obsessed with the brain.

I consider one of the most problematic aspects of such a neuroculture to be neuroessentialism. As Adina Roskies has stated, "many of us overtly or covertly believe in a kind of "neuroessentialism," many people think that our brains define who we are, even more than our genes do. Thereby, in investigating the brain, we investigate the self" (Roskies 2002: 22). Peter Reiner offers another definition of neuroessentialism, describing it as the position according to which "when we conceive of ourselves, when we think of who we are as beings interacting with the world, the *we* that we think of primarily resides in our brain" (Reiner 2011: 116).

M. García-Granero (✉)
Universidad de Valencia, Valencia, Spain
e-mail: marina.garcia-granero@uv.es

P. Calvo, J. Gracia-Calandín (eds.), *Moral Neuroeducation for a Democratic and Pluralistic Society*, https://doi.org/10.1007/978-3-030-22562-9_8

When consulting the neuroscientific literature, it may seem as if by studying the brain, both neurophilosophers and neuroscientists are believed to be studying the self as well, and some of them have asserted that we are, indeed, our brains. For instance, we find striking titles and expressions, such as *Touching a Nerve. The self as Brain* (Churchland 2013), "How Does the Brain Know When it is Right" (Singer 2005), or the *Neuronal Man* (Changeux 1997). The brain is presented as the subject, not only of neuroscientific research, but as the agent that performs actions and/or holds beliefs and thoughts, as if the brain were able to give an account of every single aspect of human experience. One could argue that these expressions are hyperboles whose sole goal is to attract the attention of potential readers. Nonetheless, they still contribute to the spread of a reductive understanding of the self and a narrow notion of what life is.

At the same time, it should be noted that many neuroscientific studies have tried to explain philosophical notions such as the self, human dignity, identity, subjectivity or agency, as well as questions regarding the meaning of life. Among neuroscientists there is enormous interest in finding the neural basis of ethical concepts such as personhood, which for example is described as "the product of an evolved brain system that projects itself automatically onto the world whenever triggered by stimulus features such as a human life-face, body or contingent patterns of behaviour" (Farah and Heberlein 2007: 37). As can be noted, the content of these explanations and definitions is usually merely biological and it reduces philosophical concepts and social phenomena to individual brain entities and neural processes.

The proposal set out here consists of maintaining a relationship between philosophy and neuroscience that does not lead to the substitution of the former by the latter, but instead to a mutual collaboration in which every discipline maintains its specificity. Our epistemic goal is to grasp the full complexity of both disciplines and not for the concerns of one to be reduced to those of the other. Essentially, philosophical analysis must be combined with scientific enquiry into the nervous system. Recently, Evers, Salles and Farisco (2017) highlighted the common tendency to underestimate the role and value of conceptual analysis. The task of this chapter is to follow up on their critique and subject the meaning and scope of neuroessentialism, along with its ethical challenges and implications, to a searching analysis, which in turn provides a path to replace neuroessentialism with a notion of subjectivity that eschews reductionism.

In so doing, the discussion that follows adheres to Javier Gracia-Calandín's suspicion that the neuroscientific explanation of the world may well become a neuromyth (Gracia-Calandín forth). A key facet of this neuromythical world-view is neuroessentialism. The thesis set out in this chapter is that neuroessentialism not only reduces subjectivity to the brain, and overlooks the rest of the body, but it also casts aside aspects of the social environment and the lifeworld that actually leave a huge imprint on the construction and education of the self. The dimensions of the lifeworld cannot be translated into neurological terms. Each person is a vector resulting from the nervous system, their body as a whole, and the lifeworld.

Naturally, this leads to a more holistic sense of subjectivity. Subjectivity is used here in a broadly inclusive sense that encompasses feelings, affectivity, intentionality and "all of the traits of human intelligence and experiences that have seemed resistant to scientific explanation" (Murillo et al. 2016: 2). This approach to subjectivity eschews reductionism thanks to its attention to the emergent characteristics of the self as a dynamic and purposive agent.

The discussion that follows is divided into three sections. The next section studies a set of concepts from Friedrich Nietzsche's philosophy that provide valuable insights into how to maintain a critical outlook towards the study of moral neuroeducation. I argue that Nietzsche provides a philosophical support and that his thought anticipated some of the critiques discussed later in the chapter. The following section focuses on the concept of embodiment, or bodily subjectivity, and proves that the body of a person is by no means irrelevant, and that instead it participates both in cognition self-consciousness. The final section takes up the philosophical concept of the lifeworld in order to establish and articulate a bodily notion of subjectivity that is situated and influenced as a being-in-the-world.

8.2 The Vigour of the Nietzschean Critique of Science from the Point of View of Value

Nietzschean perspectivism allows us to identify the study of the body as a real hermeneutic phenomenon. Nietzsche's philosophy works beyond the division of biological and mental phenomena and overcomes both naturalism and idealism, because the primordial centre of his attention is interpretation. It is also invaluable for the question at hand, insofar as it combines three key elements: a critique of positivism, a defence of bodily subjectivity and the Nietzschean notion of perspectivism.

Nietzsche's perspectivism emerged as a critique of positivism, metaphysics and religion. He was not a philosopher of science, but first and foremost a critic of culture who questioned the *problem* of science (Babich 2008), and "the victory of the scientific method over science" (Nietzsche 2009; Posthumous Fragment [=PF] 1888 15[51]). He highlighted the necessarily interested and evaluative character of all knowledge. Even the natural sciences respond to a host of interests, necessities and conditions related to the survival of the human being as a species. As such, the error of positivism is not one of content, but methodology: it denies the nature, structure and function of perspectives, which are a basic condition of life. Knowledge is never selfless; science, too, stems from a will to power and dominion. According to Nietzsche, perspectives fulfil an affirmative function in so far as they nurture interpretations and are a condition of our experience of the world. We cannot know things as they are in themselves, nor independently of our interpretative appropriation: "There are no "facts-in-themselves", *for a sense must always be projected into them before there can be "facts"*. The question "what is that?" is an imposition of meaning from some other viewpoint. (…). At the bottom of it there always lies "what is that for me?" (for us, for all that lives, etc.)" (PF 1885 2[149]). Every approach to the world and experience of reality is tied to a specific perspective, or to a set of parameters that enable a concrete experience of the world. These parameters establish that seeing is in fact seeing something, because our point of view orients our perception in a particular direction, perimeter and horizon.

Nietzsche highlights the question of value, which is more fundamental than that of truth, and in fact, truth is a value that is maintained because of the faith we invest in it. It is the value of certain knowledge that must be sought after. This critical inquiry from the optic of life is not ontological, it does not question the existence of an object, but instead it aims to elucidate why the said object should, or should not, be important to us. The idea that there are no facts, but only interpretations, is first and foremost an exercise of intellectual honesty, an acknowledgement of the experiential limits of our intellect and a critique of positivism. But perspectivism does not entail relativism, because even if we can elaborate an infinity of interpretations regarding a story or an event, not all of them are worth something.

The point of view provided by consciousness has increasingly prevailed both in science and culture, whereby the bodily capabilities of creating meaning have been overshadowed. This ascetic situation has had, as a consequence, a disdain of the bodily self, when, in fact, it is the body that makes both knowledge and the judgement of value possible. The body performs "an intellectual activity that does not enter consciousness, as pain does as a consequence of a wound. Probably an inner event corresponds to each organic function; hence assimilation, rejection, growth, etc." (PF 1885 40[15]). The correct representation of our subjectivity is bodily, and it should be noted that Nietzsche speaks of the body as the living organism of the naturalists (*Leib*), rather than the body as the lifeless and calculable entity of physicists (*Körper*). The living body is not just a physical entity, it is actually the incarnation (*Einverleibung*) of a person. In particular, there are spheres of vital experience, such as individuality, intimacy and privacy, that concern corporeity. On the other hand, the experimental sciences homogenise (bodily) experiences making them equivalent, which is of course a necessary step to carry out experiments, but life cannot be comprehended with the rigour of science, nor with the logic of knowledge. Through a translation of this sort, perspectivism is reduced to unilaterality and even dogmatism.

Nietzsche acknowledges that the starting point of our individual body is always located in a particular angle within each situation. Living is a condition of knowledge, and the knowledge that we are capable of is obtained thanks to the body. As opposed to the Platonic tradition that understands the severance with the body as a requisite of true knowledge, Nietzsche argues that the body is the starting point of knowledge. However, that knowledge is not absolute and does not exist in itself. It is perspectivist and limited, and there is no escape from it, nor any other rational way to access a so-called "real" world. The true instinct of life is situated in corporeality in the most unconditional manner (*The Genealogy of Morals* III §12), it is the instance with the greatest potential of probity from the point of view of life.

The vital experience of the body is the fundamental phenomenon from which interpretations arise. Nietzsche's genealogical critique reveals the experiential conditions of truth, that should be transvalued as a vital function by following the body's conductive thread. Every value judgement is itself a production of meaning that is mediated physiologically, historically and culturally. Thereby, an interpretation is not an explanation, but an introduction of meaning and a basic figure of thought.

Genealogy places physiology as the master of all questions, because the physiological component is decisive for the interpretation of all vital symptoms, the value of all values and their repercussion on people's development. Nonetheless, the knowledge provided by science should not be understood in a positivist manner, but in a hermeneutical frame of reference, because all natural phenomena are comprehended as intelligent processes of interpretation that are susceptible to change. Nietzsche pursues a project of reconciliation between nature and culture without ascribing a privileged position to the natural sciences. He repeatedly insists on the "most delicate question": whether science can provide us with the goal of actions (*The Gay Science* §7). Science cannot be our guide when it comes to making moral decisions, because a naturalist perspective can neither create value, nor elucidate the value of values. Scientific advances can be used for different purposes, towards freedom or domination, towards equality or inequality, and that meaning or purpose is what will make them either valuable or worthless. One of the key tasks of moral education is to cultivate the faculties that are necessary to discern value. Perspectivism does not entail that we cannot choose between a plurality of points of views. Not all values, and not all forms of life, are worth the same, nor are they equivalent in their manner of introducing meaning into life: in fact, "our strength has to manifest itself in the way we *choose*; we must be judges" (PF 1877 23[85]).

Perspectivism cannot be equated to scepticism, nor does it seek to disregard science. Instead it opens a space for the transvaluation of scientific elements and the interpretations that we can create as a result of scientific discoveries. In fact, Nietzsche was very much aware of the scientific context and literature of his day, and his hermeneutical genealogy incorporated a neurological component: it paid special attention to theories of pain, and also analysed how the superior functions of the spirit connect with organic processes upon which the brain exerts its particular function. Abraham Olivier has highlighted this neurological component in Nietzsche's thought and suggested that it marks the origin of a neurophilosophy (Olivier 2003).

In my opinion Nietzsche's greatest contribution to neurophilosophy would be his critical perspective. He did not yield to reductionism. He defended the view that the totality of an organism thinks, and that all of the organs play a part in the process of thinking, feeling and willing. The brain is only a "great apparatus of centralisation", because all the mental processes are moulded by the organic processes of the body. The brain is responsible for our perceptions and images, but it is also, in itself, a part of these representations. Just like the eye does not see, but we see, the brain does not think, it just enables thinking, *we* think. In fact, we are able to think about our brain, because we create a cultural representation of the physical entity. That is why the brain in itself "possesses absolutely no reality" (PF 1873 27[37], 27[77]). It was not in vain that Nietzsche argued that the whole of the body is a "great reason" and that all organic life is coordinated with a spiritual happening. He encouraged his readers to analyse the interpretations that are practised by humanity in order to appreciate their value, or lack thereof, in which case we should create other interpretations, ones that allow us to increase the value of the life they bring about. This genealogical task is a preamble that enables educational achievement.

How do these questions apply to contemporary neuroethics? As I will show in the rest of the chapter, the main challenge raised by some circles of neuroscience, is the imposition of a particular interpretation of the world (that is, neuroessentialism) as a scientific "fact" and as the only scientifically accurate way to conceive of our subjectivity and self-understanding. Neuroessentialism is not a scientific knowledge, it is a particular interpretation of the world. Neuroscience takes advantage of its cultural status to spread a world-view whose validity cannot be judged by science, only by ethics and philosophy. Furthermore, not only should we question whether the viewpoint of neuroessentialism is true or false, we should also ask ourselves what value it has. We should question the value of its imprint on reality and ethical life: whether it enriches our self-experience and whether it improves our understanding of ethical issues such as personhood or responsibility. Does neuroessentialism contribute to the creation of new springs of meaning, or does it promote nihilism? On the other hand, are we able to create and *choose* other interpretations that could introduce more meaning into life?

8.3 Embodiment, Emotions and the Sense of Self

As we have seen in the previous section, Nietzsche provides a precursor for a sustained critique of positivism and neuroessentialism, while also advocating a bodily concept of subjectivity. In this section, I will argue that the Nietzschean critique continues to be topical in the context of embodied theories developed in cognitive science and neuroscience.

Embodied theories of cognition defend that the bodily systems that have evolved for perception, action and emotion, contribute to a higher cognitive process. The close connection between the body and education is being studied by cognitive science and neuroeducation in order to elaborate new educational policies based on the mind-body connection. Indeed, educational strategies based on the embodiment approach developed in cognitive science have provided good results both in mathematics, using embodied mechanics and action in the form of gesture, and language acquisition and reading comprehension, as shown by the indexical hypothesis (Glenberg 2008). Language and thought are inextricably shaped by embodied action. Human reason and learning are made possible thanks to the mechanisms of the body. Embodiment approaches to learning have shown that the most overlooked aspect of human intelligence is that it is embodied and requires having a physical body that interacts with the environment through its bodily constraints. The functional design of a sensorimotor system has an important impact on tasks like concept formation and reasoning. Our sense organs are dynamic instruments of exploration and play a central role in our creation of an understanding of the world

(Anderson 2004). According to Lindblom, the emerging viewpoint of embodied cognition is that "cognitive processes depend on experiences that come from having a body with particular sensorimotor capabilities interacting with the surrounding world" (Lindblom 2015: 82). Or in other words, the situatedness of the body provides a perspective. Therefore, we should emphasise the interaction between agents and their environment. Ownership and agency, likewise, require a body acting in time and space and are meaningless without it (Damasio 2000: 145).

Antonio Damasio, one of the world's foremost neurologists, has repeatedly highlighted that the brain cannot be studied in isolation. Mental states are actually deeply embodied, and each brain activity should be regarded as a part of a complex interplay between the organism and its environment. According to Damasio, not recognising that mind, body and environment reciprocally shape each other was Descartes' error (Damasio 1994). The brain alone is not a proper unit of analysis, because the interaction between the mind and the environment is made possible thanks to the organism. For example, perceiving the environment is not just a matter of having the brain receive direct signals from a given stimulus, the organism acts on the environment as well. The mind is derived from the entire organism as an ensemble. In fact, even partial blocking of brain-body traffic causes changes to an individual's state of mind, for example in patients with a spinal cord injury (Ibid: 227).

The physiology of emotions and feeling is part of the mechanisms that regulate life. Any study regarding emotion and feelings is linked to the question of the body, its impulses and its motivations. Emotions and feelings play important roles throughout the organism. According to Damasio, emotions are played out under the control of both the subcortical and neocortical structures, and they are the most elemental products of the body because they are a part of an evaluating dynamism at the neurophysiological level. On the other hand, feelings offer us the cognition of our visceral and musculoskeletal state, they let us "mind the body" (Ibid.). Feelings are just as cognitive as any other perceptual image and depend on cerebral cortex processing (Ibid: 159). They are an essential feature of intuition, cognitive guidance and decision-making, or in other words for the practical use of reason and moral judgement.

Moreover, there is no such thing as a pure perception of an object within a sensory channel, for instance, vision. Perception also contains signals deriving from emotional responses to a particular object. Damasio poses the example of a fast-approaching car, which causes fear (Damasio 2000: 145). The perception is accompanied by both musculoskeletal and emotional signals, albeit in different proportions. They contribute to the organism processing its perception of the object in a satisfactory manner. The organism is involved in relating to a certain object, and that object within the relation causes a change in the organism (Ibid: 133).

Somatic markers are "a special instance of feelings generated from secondary emotions [that] have been connected by learning, to predicted future outcomes of certain scenarios" (Damasio 1994: 174). The critical neural system for the acquisition of somatic marker signalling is located in the prefrontal cortices. Most somatic markers are acquired by experience and created during the process of education and socialisation in accordance with the social conventions and ethical rules of the culture to which the person belongs. Somatic markers constitute a system or repertory of automatically qualified predictions that connect specific classes of stimuli with specific classes of somatic states. Thereby, they create an evaluation that anticipates the future. As an adaptive behaviour, it forces attention onto the negative outcome to which a given action may lead. The decisive element is the type of somatic state and feeling produced by a given situation at a given point in the person's history. Yet Damasio himself warns us that a somatic marker may not be sufficient for normal decision-making, since a subsequent process of reasoning and final selection should still take place (Ibid: 173). Somatic markers do not deliberate for us —deliberation being the properly moral task—, but they do assist the process of deliberation. The example of somatic markers is key because it illustrates the deep imbrication of the nervous system and body. "A mind is so closely shaped by the body and destined to serve it that only one mind could possibly arise in it. No body, never mind. For any body, never more than one mind" (Damasio 2000: 143). Given the ways in which the body shapes perception, a different body would give rise to a different perception.

Thanks to advances made in neuroscience the issue of value can also be explored from a neurological perspective and the origin of the value judgements employed in regulating life can be studied in terms of homeostasis. The role of value in regulating life seems to be fundamental. Damasio distinguishes between basic and sociocultural homeostasis. Basic homeostasis is an adaptive behaviour that operates even within unicellular living creatures. Their homeostatic feelings intervene in the solution of the essential problems involved in regulating life, whereas cultural homeostasis is adjusted within every individual through experiences related to sociocultural circumstances, as in the case for humans. The result is the shaping of affect-related phenomena such as social cooperation, in-group and out-group dynamics, cultural identities that are built as a result of factors like past experiences or related historical and geographic factors, and the deployment of social emotions, such as compassion, altruism, gratitude, and indignation (Damasio and Damasio 2016: 127).

The biological assessment of value is an activity according to which the brain unconsciously ponders the benefits and disadvantages of an action. Consideration of this perspective reveals a biological root for the ethical notion of value, which plays a role in the regulation of homeostasis and the vital needs involved in the struggle for survival and the pursuit of well-being. Damasio's research verifies the real connection between reason, emotion, feelings and the process of valuation. "You simply cannot escape the affection of our organism, motor and emotional most of all, that is part and parcel of having a mind" (Damasio 2000: 148). In fact, when we recall an object, we don't just remember sensory characteristics of the actual object, but also the accompanying motor and emotional data, the past reactions of our organism to that object. A value judgement is not only an expression of thought, but also an expression of preference, which is the result of an emotion.

Nonetheless, as Jesús Conill has pointed out, one of the unresolved tensions of this explanation is the notion of life that it disseminates (Conill 2013: 92). Damasio's scientific contribution is fundamental and extremely valuable, and he does not try to reduce ethics to biology. There is always a place for the moral point of view, which transcends the interests of an immediate group and even that of the species (Damasio 2010). Yet, there is also a need for an in-depth philosophical reflection on the question of the meaning and value of life. If the only important criteria for homeostasis are reproduction, survival, and well-being, then there is an important hermeneutic and axiological deficit since all of these can be reduced to utility. The moral duty of a person is not only to survive, but to lead a good life according to the principles of justice and to pursue the vital project that each one of us has reasons to esteem. Morality presupposes an emotional system and the capacity for preferential selection. The conservative tendency towards survival and well-being reflects without a doubt the most basic biological value, but sanctions of this kind are merely causal motivators and carry no real moral justificatory force.

Scientific observations do not dictate how they are to be interpreted, that is why attention should be paid to the interpretation of education and the idea of a moral life that is being widely disseminated (Conill 2019). Neuroscientific experiments use indirect measures of the structure and function of the brain that are laden with value judgements and assumptions, which are often shaped by societal, political and economic factors. The mainstream of so-called neurophilosophy pretends to disseminate a strictly scientific world-view, as if such a thing were possible, but does not acknowledge how this is embedded with value judgements. Instead, it actually disseminates a particular interpretation of reality that influences the symbolic universe of today's world.

Studies around neuroplasticity have proved that such a reductionist vision is inadequate (Draganski et al. 2004). The connections between neurons are not prespecified in the genes. Learning, training and experience influence the brain's development within the boundaries of a "genetic envelope" (Changeux et al. 1973). Furthermore, the biological sciences are concerned with open systems that exchange mass and energy with their surroundings or environment, which is why the analysis of the vital context and the lifeworld should be essential for the work of moral neuroeducation, neuroethics and neuroscience in general. The past history of individuals within their environmental context and their ongoing dynamic interaction contribute to what and how they are today and explains the cultural differences of their embodied minds. The participant's involvement in shared lifeworld practices needs to be taken into consideration when interpreting experimental results. Accordingly, this would be a non-reductionist use of neuroscience. In fact, a growing number of researchers in neuroethics and neuroscience are defending the view that philosophy should not accept the findings of neuroscience without reflecting on the assumptions that underpin scientific analysis, and as a fruit of this stance, "non-reductive science" is no longer dismissed as an oxymoron (Murillo et al. 2016).

It is worth remembering that normative judgements cannot be logically derived from descriptive statements, and this is a conceptual mistake that one usually encounters in neuroscientific literature. To posit a direct link between descriptive considerations derived from observations about the brain and the moral justification of social norms is to commit a naturalistic fallacy. The case of aporophobia is especially helpful to illustrate this point. The fact that we are neurologically predisposed to xenophobia and aporophobia does not morally legitimate such predisposition.

Instead, it is our moral duty to overcome it, and to cultivate a moral disposition of respect, recognition and cooperation (Cortina 2017; Pallarés-Domínguez 2019).

The brain is, without a doubt, an essential centre of our mental life and is a necessary condition for our sense of personal identity, but it must not be confused with the person. The brain is the basis of the mind, but the mind does not identify itself with the brain. Neuroscientific studies explore the biological basis of what we are as human beings, but they cannot account for our daily experiences as persons and citizens of modern societies. A human being does not become a person thanks to biology, but through education. Furthermore, brains are not held accountable for their actions, people are (Cortina 2011: Chap. 8). Therefore, neuroessentialism is proved to be worthless when it comes to pondering and judging the responsibility of a person. Instead, the irreducible feature of personhood must guide the analysis of issues such as personal identity and responsibility.

We should extend our analysis to all the processes and mechanisms that contribute to cognition, rather than limiting it to those processes and mechanisms that take place within the skull. A critical aspect of self-awareness is the sense of being situated within the limits of our own body. Sensations facilitate a direct experience of the living body, one related to our own existence, and self-awareness is not independent of our somatic perceptions, because primordial representations of the body proper provide a natural reference for what happens to the organism, anchoring the self that exists at any moment (Damasio 1994: 235). The representation of a body constitutes a basis for the concept of self, "much as a collection of representations of shape, size, colour, texture, and taste can constitute the basis for the concept of orange" (Ibid: 240). It is true that a person can even imagine themselves many years ahead with a different bodily appearance, and this is because representations of the body landscape "as if" it had changed are created directly within sensory body maps that are controlled by neural sites such as the prefrontal cortices (Damasio 2000: 80).

It is also true that most bodily processes function at an unconscious level, for example, I do not have to consciously think about blinking. The focus of our attention is usually concerned with representing external events, where it is most needed for adaptive behaviour, but that does not mean that the body's representation is absent. According to Damasio, brain processes that regulate the body and map body signals implement the proto-self. However, the proto-self does not occur in one place only, but instead "emerges dynamically and continuously out of multifarious interacting signals that span varied orders of the nervous system" (Damasio 2000: 154). The proto-self should not be confused with the rich sense of self. We are not conscious of our proto-self, as it holds no knowledge and language is not a part of its structure. The proto-self is "a coherent collection of neural patterns which map, moment by moment, the state of the physical structure of the organism in its many dimensions" (Ibid.). The organism, taken as a unit, is mapped in the brain within structures that regulate the organism. It is a first-order representation of a current body state, of which we are not conscious, a reference point at each point of its presence.

On the other hand, consciousness only arises when an organism, an object, and its relationship, can be represented. Consciousness occurs when we map the relationship between ourselves and an object, or the rest of the world, or when we think about the bodily processes happening in our organism and the changes that are caused by external objects. The consciousness of a body as our own is precisely what enables us to understand that our conscious life is rooted in an organic substratum and provides our sense of emplacement in the world. At the conscious level, Damasio distinguishes between two kinds of self. Firstly, the core self and the autobiographical self. The core self can be triggered by any object and is a second-order non-verbal account that occurs whenever an object modifies the proto-self. Secondly, the autobiographical self requires the core self to begin its gradual development, because it is based on permanent, yet dispositional records of core self experiences that are turned into explicit images. The autobiographical self is based on memories of multiple perspectives on individual experiences of the past, the present of the anticipated future (Damasio 2000: 174–175). These sets of memories describe the identity of a person. According to this line of thought, the body can be conceptualised as the centre of coordinated vital activities, including the generation of consciousness, presence and temporality (Murillo 2016).

The brain gives rise to consciousness by enabling an exchange between the person and the world, which depends greatly on where we are and what we do. In order to understand human consciousness, one should not focus on the brain alone, but on the brain in the context of a living body and the setting provided by a person's active life. There is a need for alternative world-views that do not sever the brain, body and world from one another. The opposition of mind versus body is a false dilemma, because the tacitly assumed disjunction is false (Klima 2016).

In regard to this it is worth mentioning that research on Artificial Intelligence has concluded that a disembodied theory of mind is biologically implausible and that being embodied is a core property for intelligence both in humans and artificial systems. Therefore, proponents of New AI, such as Rodney Brooks, have focused on the challenges of getting robots to experience the world through their interaction with the environment, and for them to receive immediate feedback on their own sensations (Brooks 1999). This offers proof that the physical embodiment and material cause of a cognitive agent is neither incidental nor trivial, and even cognition that takes places "offline" is based on sensorimotor mechanisms.

Nonetheless, it is my view that in no way can we equate human cognition to AI cognition, because humanoid robots do not experience the world in the same way as people, and mechanistic theories of AI neglect the subjective nature of the living body. The German concept of *Leib* as a notion of phenomenological embodiment allows us to stress the essential part of being a living body situated in a subjective world.

The traditional computational paradigm in cognition science and neuroscience defends the view that cognition takes place inside the skull in the form of abstract symbol manipulation, a view according to which the body would only serve as an input and output device. However, as has been discussed, the embodied account of cognition provides an alternative to this view and it has been argued that cognition is an activity situated in real-time that is fully interwoven with perception and action. The social nature of cognition should also be stressed, because cognition "is what occurs when the body engages in the physical, cultural world and must be studied in terms of the dynamical interactions between people and environment" (Gibbs 2006: 9). An adequate understanding of a person's behaviour would seem to depend on the dynamic organisation of their body as a whole and how it both influences and is influenced by the purposeful activity of the individual within their environment.

8.4 The Socially Situated Self-Experience in the Lifeworld

It is now necessary to explain in more detail how the relationships between the physical, social, moral and political worlds arise, whereby the embodiment thesis needs to be developed beyond the body towards a social dimension. In my view moral neuroeducation should incorporate the perspective of the self's-experience within the lifeworld and adopt a non-reductive and empirically committed approach to education, ethics, subjectivity and life.

The "Lifeworld" is a philosophical concept developed by Husserl in order to designate the vast space of experiences, intersubjective relationships and values that are familiar to us in the day-to-day interaction with people and things. It is in the lifeworld where value judgements are generated, as well as the experiences that comply with them. In *The crisis of European sciences*, Husserl criticised the restriction of world-views to logically integrated systems, whose assertions can only be formulated quantitatively (Husserl 1993). Habermas, for his part, presents the "lifeworld" as the historical and empirical place of emancipation and social critique. When we contemplate a society as a lifeworld, we are able to question its normative structures, values and institutions; that is, we highlight the horizon upon which citizens participate and involve themselves in comprehension processes. According to Habermas, the guiding principles of the lifeworld are understanding and solidarity, which are principles that underlie the practices that coordinate our actions, that produce and reproduce life, and enable the transmission of values (Habermas 1997).

The brain's functional architecture grounds moral agency (Richart 2019). Therefore, we must ponder the brain's primordial functioning both at the core of the human body, and also within a concrete lifeworld, that is, within the democratic and pluralistic societies we wish to build. For example, moral neuroeducation involves the education of values because moral conflicts are not biological, but of a social nature. Many of the moral conflicts that happen in our societies involve a clash of values and interests. That is why one of the most important tasks of moral neuroeducation is to contribute to the solution of these conflicts by considering both neuroscientific findings and the social context of the lifeworld. Specifically, the method of critical hermeneutics allows us to propose specific processes for the application of moral knowledge based on specific contexts for all human activities (Conill 2016), as in the case of ethical economy (Calvo 2018).

As has already been hinted at in the previous section, organisms are embedded in an environment and are deeply affected by their dynamic interaction with other organisms, society and surroundings. In fact, a person "is not a self-contained module or an autonomous whole. We are not like the berry that can be easily plucked, but rather like the plant itself, rooted in the earth and enmeshed in the brambles" (Noë 2009: 69). People are not just brains, but neither are they mere bundles of bodily experiences. We should not be bound to a viewpoint of the brain and body, because cognition and human experience are influenced and constrained by a real-time interaction with the world around them. The dynamic nature of embodiment is in fact a result of its sense of being located within the lifeworld: "The self cannot be reduced to the brain alone, but neither is it limited by the boundaries of the body, because there can be no self without others" (Stapleton and Froese 2016: 125). It may be proposed that there is a biological basis for meaning that can help us understand the link between the biological substratum of the brain and the cultural dimension of interpretation, which may be understood as part of "the neurohermeneutic process where the cultural memory networks are activated and retrieve the perceptual and procedural attributes of cultural signs associated with sensations" (Reyna 2006: 137). This process also reveals the importance of shared experience within the lifeworld; it provides a landscape for civil society and social interaction. The experience of self involves pragmatic and social interactions in the lifeworld, which is where people engage with their surroundings and construct meaning.

Human meaning, understanding and rationality are the results of the embodied experience of a person in the lifeworld. As Shaun Gallagher has stated:

> (...) mental states do not fly through thin air between minds; nor are they simply replicated in matching brains or externalized in pure speech acts. Rather, human feelings, intentions, thoughts, and beliefs are deeply embedded in backgrounds and contexts, in embodied social interactions and communicative practices in the everyday lifeworld, and all of these phenomena are characterised by a great many differences that need to be recognized and acknowledged (Gallagher 2012: 104).

The meaning of a norm that is to be applied cannot be disconnected from the context of the activity, such as the physical situation and setting, the roles played by individuals, etc., which are what give the norm meaning. Meaning needs to be shared and recognised by others in order to be considered appropriate, and to continue to be meaningful and effective in its communicative task. That is why a complete explanation of social cognition "is not possible simply in terms of neuronal processes; since it clearly involves others in extra-neural contexts and worldly events, it is not reducible to the worldless realm of brain events" (Ibid: 96). The richness of human experience, informed by emotion, memory and imagination is not reducible to brain events. We cannot translate the experience of the lifeworld into brain events. Nevertheless, the experience of the lifeworld is actually necessary to understand many neurological disorders. This view is clear when it comes to the example of mental illness, which should be analysed and treated, "on the one hand, as inseparable from the living organism and on the other, as inseparable from the patient's lifeworld or social environment" (Fuchs 2012: 331). Other biological phenomena such as self-consciousness or self-understanding are likewise the result of the connections between the brain, the human body and the cultural programmes reproduced through social communication. Thanks to experience within the lifeworld, we come to understand how contexts can inform the emotions, intentions, and thoughts of others, and we are able to create a theory of mind.

An integral notion of human reality such as this addresses our involvement with the world through language, the tools we use, and the social relationships and the practices we develop. To learn is to modify the brain tissue as a result of experience and a people's thoughts and actions leave traces in their neural system. The brain is a very malleable structure that is affected by everything that happens outside of it. By changing how we act and through learning, we can change our own form, body and mind (Noë 2009). It is understood that a person's place within the lifeworld also affects and progressively changes the way their neural network is configured (Gillett 2009). The neural operations that construct life-worlds are learned and not prescribed by our genes, which is why different groups and cultures build different life-worlds.

According to Habermas's theory of communicative action, there are two kinds of language and modes to approach and tackle the world. The first perspective is that of the observer of objects and events. The second is the perspective of the participant in communicative actions involving other subjects who are likewise participants. In the first case, a subject observes objects; for example, in the case of brain images and neuroscience researchers aim is to produce science and attain objective affirmations. In the second case, different people recognise each other as valid interlocutors, thereby revealing the existence of intersubjectivity (Habermas 1997). In a similar vein, Sturma expresses that:

> (…) in the perspective of the perceiving, experiencing and acting person, the everyday world and brain research do not converge. We are always either on the side of self-consciousness and experience, or on the side of neuroscience, studying neural processes from the outside (Sturma 2016: 15).

Therefore, we can distinguish two distinctive perspectives, which need to both coexist and be conserved, in regard to which a series of points must be made. Firstly, it is not possible to separate the interpretations an individual creates about the world from the experience that is needed to access such a world, neither can we objectify the comprehension that can only be acquired through participation in the world. Secondly, the instrumental perspective of the scientific approach has debilitating effects when it is used to analyse moral, political and social issues. The extension of an objectifying third-person perspective to the concerns of the first-person lifeworld is devoid of meaning and, therefore, promotes nihilism. No objective third-person description can fully capture the perspective of a first-person view, and to live through the experience of another is a logical impossibility. In regard to this, Walter Glannon provides an illuminating example to prove that there will always be an explanatory gap between the objectivity of the brain and the subjectivity of experience:

> Even if the fMRI investigator provided the relevant stimuli to me in the scanner, and all of the brain regions (…) were activated, this would not tell him or her what it was like for me to be in Andalusia or to recall it. (…) How my brain records these memories at a lower level, and how my mind constructs meaning from them at a higher level, is a function of unique contextual features (Glannon 2011: 18–19).

When neuroscientists aim to study observable phenomena, they are physically unable to include the background of the intersubjectively shared lifeworld. While this should not pose a problem for scientific research, it reveals frontiers of knowledge that it is impossible for neuroscience to cross. Both perspectives are needed, each with its own specificity and domain, in order for both science and philosophy to create valuable knowledge together. Without consideration of the intersubjectivity of the act of understanding, which is typical of the lifeworld, there is no objectivity of knowledge.

For example, the self may be a construction, but it is not an illusion. People are connected to cultural programmes that are reproduced through social communication between speakers, hearers and bystanders, for example the use of a language is meaningful thanks to the existence of a social practice. Both verbal and bodily language are examples of the phenomena of a lifeworld that is intersubjectively shared. The role of the body in social interaction, or so-called social embodiment, highlights how corporeal states such as postures, arms movements and facial expressions play important roles in processing social information as well as interactions.

Critical theory does not originate from the perspective of the external observer, but instead, from the perspective of the lifeworld and its real parameters. Members of the lifeworld are not merely linguistic subjects; they are also affected positively or negatively by all systemic ambits (the state and the economy) and other configurations of practices that are external to individuals but still affect them. Habermas termed this idea as the "colonization of the lifeworld" (Habermas 1997). That being said, we should not forget that the lifeworld is not immune to the influence of structures of domination and oppression. There is a need to analyse the normative structures, values and institutions of the lifeworld from a moral perspective (García-Granero and Ortega Esquembre 2019). The historicity of the lifeworld should be constantly stressed, due to the existence of social norms that are blatantly unfair and oppressive, and that these norms, in turn, also contribute to shaping the way the world is interpreted from a first-person perspective. We should be able to give an account of how the lifeworld comes to be established and whether prevailing norms and world-views are actually valid and able to count on the recognition of all the people affected by them.

8.5 Conclusion

In this chapter, Nietzschean perspectivism and criticism has been invoked in order to question the value of neuroessentialism for democratic and pluralistic societies, and whether they help to create a meaningful notion of subjectivity. We conclude that it does not and can, in fact, promote a nihilistic world-view. Instead, an integral and richer notion of subjectivity is more valuable, because it allows us to establish and respect the irreducible character of subjective experience in the lifeworld, where norms, actions and interactions obtain their meaning.

From modern philosophy to contemporary neuroethics, the view of the self and the mind as being embodied does not only come from philosophical analysis, but also from empirical evidence. Subjectivity cannot be confined to the brain. Instead, neuroeducation, neuroethics and neuroscience should acknowledge both the embodiment of the mind and of consciousness. Mental states are not just brain states: their content and subjective quality is influenced by bodily systems outside of the brain. However, this embodied notion of subjectivity should also be extended to the lifeworld, because that is where people engage with each other and construct meanings within their surroundings. Our self-experience in the world is the product of a circular causation consisting of bodily, social and environmental influences, which continuously interact with each other in the lifeworld.

Acknowledgements The author is a researcher supported by a FPU contract-fellowship from Spain's Ministerio de Ciencia, Innovación y Universidades (Reference: FPU15/04085).

References

Anderson, Michael L. 2004. Embodied cognition: A field guide. *Artificial Intelligence* 149: 91–130.

Babich, Babette. 2008. "Un problema con cuernos… El problema de la ciencia misma". La crítica de Nietzsche a la razón científica. *Estudios Nietzsche* 8: 13–52.

Brooks, Rodney A. 1999. *Cambrian intelligence: The early history of the new AI*. Cambridge, MA: MIT Press.

Calvo, Patrici. 2018. *The cordial economy – Ethics, recognition and reciprocity*. Cham: Springer.

Changeux, Jean-Pierre. 1997. *Neuronal man. The biology of mind*. Princeton: Princeton University Press.

Changeux, Jean-Pierre, Philippe Courrège, and Antoine Danchin. 1973. A theory of the epigenesis of neuronal networks by selective stabilization of synapses. *Proceedings of the National Academy of Sciences of the United States of America* 70 (10): 2974–2978.

Churchland, Patricia S. 2013. *Touching a nerve. The self as brain*. New York/London: W. W. Norton & Company.

Codina Felip, María José. 2015. *Neuroeducación en virtudes cordiales. Cómo reconciliar lo que decimos con lo que hacemos*. Oviedo: Octaedro.

Conill, Jesús. 2013. Neurorracionalidad práctica y valor biológico. *Daimon Revista Internacional de Filosofía* 59: 89–102.

———. 2016. The hermeneutical and deliberative nature of applied ethics. In *Public reason and applied ethics*, ed. A. Cortina, D. Garcia-Marzá, and J. Conill, 85–99. New York: Routledge.

———. 2019. Must ethics for moral neuroeducation be naturalistic? In *Moral neuroeducation for a democratic and pluralistic society*, ed. Patrici Calvo and Javier Gracia-Calandín. Cham: Springer.

Cortina, Adela. 2011. *Neuroética y Neuropolítica. Sugerencias para la Educación Moral*. Madrid: Tecnos.

———. 2017. *Aporofobia, el rechazo al pobre. Un desafío para la democracia*. Barcelona: Paidós.

Damasio, Antonio R. 1994. *Descartes' error: Emotion, reason, and the human brain*. New York: Avon Books.

———. 2000. *The feeling of what happens. Body, emotion and the making of consciousness*. London: Vintage.

———. 2010. *Self comes to mind: Constructing the conscious brain*. New York: Pantheon.

Damasio, Antonio, and Hanna Damasio. 2016. Exploring the concept of homeostasis and considering its implications for economics. *Journal of Economic Behavior & Organization* 126: 125–129.

Draganski, Bogdan, Christian Gaser, Volker Busch, Gerhard Schuierer, Ulrich Bogdhan, and Arney May. 2004. Neuroplasticity: Changes in grey matter induced by training. *Nature* 427: 311–312.

Evers, Kathinka, Arleen Salles, and Michele Farisco. 2017. Theoretical framing of neuroethics: the need for a conceptual approach. In *Debates about neuroethics: Perspectives on its development, focus, and future*, ed. E. Racine and J. Aspler, 89–107. Cham: Springer.

Farah, Martha J., and Andrea S. Heberlein. 2007. Personhood and neuroscience: Naturalizing or nihilating? *The American Journal of Bioethics* 7 (1): 37–48.

Frazzeto, Giovanni, and Suzanne Anker. 2009. Neuroculture. *Nature Reviews Neuroscience* 10: 815–821.

Fuchs, Thomas. 2012. Are mental illnesses diseases of the brain? In *Critical neuroscience: A handbook of the social and cultural contexts of neuroscience*, ed. S. Choudhury and J. Slaby, 331–344. Oxford: Blackwell Publishing.

Gallagher, Shaun. 2012. Scanning the lifeworld. Toward a critical neuroscience of action and interaction. In *Critical neuroscience: A handbook of the social and cultural contexts of neuroscience*, ed. S. Choudhury and J. Slaby, 85–110. Oxford: Blackwell Publishing.

García-Granero, Marina, and César Ortega-Esquembre. 2019. ¿Teoría crítica o inmunización del sistema? Acerca de la dicotomía habermasiana entre sistema y mundo de la vida. *Tópicos, Revista de Filosofía* 56: 311–337.

Gibbs, Raymond W., Jr. 2006. *Embodiment and cognitive science*. Cambridge: Cambridge University Press.

Gillett, Grant R. 2009. The subjective brain, identity, and neuroethics. *The American Journal of Bioethics* 9 (9): 5–13.

Glannon, Walter. 2011. *Brain, body, and mind: Neuroethics with a human face*. Oxford: Oxford University Press.

Glenberg, Arthur M. 2008. Embodiment for education. In *Handbook of cognitive science: An embodied approach*, 355–372. Amsterdam: Elsevier. https://doi.org/10.1016/B978-0-08-046616-3.00018-9.

Gracia-Calandín, Javier. Forth. *Critical neurohermeneutics*. Anuario Filosófico.

Habermas, Jürgen. 1997. *The theory of communicative action*. Cambridge: Polity.

Husserl, Edmund. 1993. *Die Krisis der europäischen Wissenschaften und die tranzendetale Phänomenologie*. Dordrecht: Kluwer Academic Publishers.

Klima, Gyula. 2016. Mind vs. body and other false dilemmas of Post-Cartesian philosophy of mind. In *Biology and subjectivity*, ed. M. García-Valdecasas, J.I. Murillo, and N.F. Barrett, 25–39. Cham: Springer.

Lindblom, Jessica. 2015. *Embodied social cognition*. Cham: Springer.

Murillo, José Ignacio. 2016. Body, time and subject. In *Biology and subjectivity*, ed. M. García-Valdecasas, J.I. Murillo, and N.F. Barrett. Cham: Springer.

Murillo, José Ignacio, Miguel García-Valdecasas, and Nathaniel F. Barrett. 2016. Biology and subjectivity: Philosophical contributions to a non-reductive neuroscience. In *Biology and subjectivity*, ed. M. García-Valdecasas, J.I. Murillo, and N.F. Barrett, 1–11. Cham: Springer.

Nietzsche, Friedrich. 2009. *Digital critical edition of the complete works and letters* (*eKGWB*), based on the critical text by G. Colli and M. Montinari, edited by Paolo D'Iorio. http://www.nietzschesource.org

Noë, Alva. 2009. *Out of our heads: Why you are not your brain, and other lessons from the biology of consciousness*. New York: Hill and Wang.

Olivier, Abraham. 2003. Nietzsche and Neurology. *Nietzsche-Studien* 32: 124–141.

Pallarés-Domínguez, Daniel. 2019. Moral neuroeducation: Proactive epigenesis and poverty. In *Moral neuroeducation for a democratic and pluralistic society*, ed. Patrici Calvo and Javier Gracia-Calandín. Cham: Springer.

Reiner, Peter B. 2011. The rise of neuroessentialism. In *Oxford handbook of neuroethics*, ed. Judy Illes and Barbara J. Sahakian, 161–175. New York: Oxford University Press.

Reyna, Steve. 2006. What is interpretation? A cultural neurohermeneutic account. *European Journal of Anthropology* 48: 131–143.

Richart, Andrés. 2019. Moral neuroeducation from a phylogenetic, ontogenetic and functional perspective. In *Moral neuroeducation for a democratic and pluralistic society*, ed. Patrici Calvo and Javier Gracia-Calandín. Cham: Springer.

Roskies, Adina. 2002. Neuroethics for the new millenium. *Neuron* 25: 21–23.

Singer, Wolf. 2005. How does the brain know when it is right? In *Neurobiology of human values*, ed. J.-P. Changeux, A.R. Damasio, W. Singer, and Y. Christen, 125–135. Cham: Springer.

Stapleton, Mog, and Tom Froese. 2016. The Enactive Philosophy of Embodiment: From Biological Foundations of Agency to the Phenomenology of Subjectivity. In *Biology and Subjectivity. Philosophical Contributions to Non-Reductive Neuroscience*, ed. M. García-Valdecasas, J. I. Murillo, and N. F. Barrett, 113–129. Cham: Springer.

Sturma, Dieter. 2016. Self-consciousness, personal identity and the challenge of neuroscience. In *Biology and subjectivity. Philosophical contributions to non-reductive neuroscience*, ed. M. García-Valdecasas, J.I. Murillo, and N.F. Barrett, 13–24. Cham: Springer.

Part III
Moral Neuroeducation in Practice

Chapter 9
Guidelines to Opening up Spaces for Shaping and Training Moral Judgement in Organizations. A Proposal Based on Neuro(Advances)

Elsa González-Esteban

9.1 Introduction

Recent advances in neuroscience have provided valuable information on the brain's neural foundations, and they have also contributed insights on how our moral judgements are formed and what elements are involved when we are confronted with ethical situations. Until the end of the twentieth century, both common knowledge and scholarship devoted to ethics prioritised rational and explicit arguments. However, since the turn of the century research on neuroethics has shown that intuition, emotions, and implicit knowledge are relevant when taking ethical decisions or designing organisations. Although other disciplines such as philosophy and moral psychology had already demonstrated this relationship, empirical evidence has now been provided to support these earlier findings, above all through the research on the brain's physiological neural foundations. (Cortina et al. 1994; Evers 2009; Farah 2002; Haidt 2001; Illes 2006; Levy 2007; Reynolds 2006). This chapter aims to identify and define organisational spaces where the moral character of an organisation can be built and forged, to do so contributions from organisational neuroscience (Becker et al. 2011; Becker and Cropanzano 2010; Cropanzano and Becker 2013; Cropanzano et al. 2017) and neuroeducation (Battro et al. 2008; Narvaez and Vaydich 2008; Narvaez 2007, 2012) are drawn on as part of an endeavour to undertake advances based on a novel critical outlook (Conill 2017; González-Esteban 2016; Lindebaum and Zundel 2013). I argue that it is within these spaces that organisations will be able to manage and specify moral emotions and moral prototypes, which can in turn be granted critical recognition and incorporated into an organisation's rationale. The creation of such spaces may also require normative guidance with regard to the processes employed for developing and training the moral judge-

E. González-Esteban (✉)
Universitat Jaume I, Castellón de la Plana, Spain
e-mail: esteban@uji.es

P. Calvo, J. Gracia-Calandín (eds.), *Moral Neuroeducation for a Democratic and Pluralistic Society*, https://doi.org/10.1007/978-3-030-22562-9_9

ments of organisations' individual members (Bowie 2009; Reynolds 2006). The proposed spaces set out below have been defined and developed within a framework of stakeholder dialogue and by drawing on proposals made within critical organisational ethics that seek to guide the management of organisations' ethical dimensions, as well as specific mechanisms such as codes of ethics and conduct, ethics committees, responsibility or sustainability reports, ethics hotlines, and ethics audits (Cortina 2011; García-Marzá and Feenstra 2013), (Freeman et al. 2010; Freeman 1984; González-Esteban 2007).[1] The discussion that follows is divided into four parts, the first of which discusses what it means to forge moral character within organisations. It is argued that this concern should be framed by an organisation's understanding of itself, which should be developed through a process of critical reflection on itself developed in response to any value conflicts that may arise when taking decisions or positions, whether these be individual or corporate. The forging of moral character is therefore addressed as an everyday occurrence and, as I will argue, it should arise from shared frameworks constructed jointly by all the actors or interlocutors involved in the activity under consideration.

Secondly, I survey the insights provided by recent studies undertaken within three growing fields of research, namely, neuroethics, neuroeducation, and organisational neuroscience. The principal aim of this survey is, firstly, to use the knowledge gained within these new fields to show how moral learning functions and, secondly, to consider the organisational implications of these insights; this knowledge can be applied as a guide or roadmap for the design of processes for developing moral judgement.

In the third part attention is turned to the identification and definition of organisational spaces that help shape and develop moral judgement in organisations; within these spaces the participation and deliberation of all stakeholders is of paramount importance. In particular, codes of ethics and conduct, ethics committees, social responsibility reports, and ethics hotlines are examined as tools organisations can apply to forge their own moral character. It is then discussed how knowledge from the *neuro* disciplines can help make these spaces and tools more efficient from an ethical point of view.

In the final part, knowledge from the neural foundations of moral judgement is drawn on to set out a series six guidelines that provide a framework to establish actions and develop spaces for participation and deliberation within an organisation. Thereby, I advocate the use of advances in the *neuro* disciplines to forge moral character in organisations in a more effective manner, yet without sacrificing any critical outlook in the process.

The theoretical approach taken in this chapter is based on the consideration of organisational ethics as an applied ethics that must operate in an interdisciplinary fashion in order to guide organisational practice. In this way organisations will be

[1]This study is part of the Scientific Research and Technological Development Project FFI2016-76753-C2-2-P, financed by the Ministerio de Economía y Competitividad de España, and part of the work undertaken by the research group of excellence PROMETEOII/2014/082 funded by the Generalitat Valenciana.

able to design their own structures and make the best decisions according to three core concerns: justice, prudence and responsibility (Cortina et al. 2008; Cortina 2013). Recent advances in the fields of neuroethics, neuroeducation and organisational neuroscience have provided invaluable information about the neural foundations of moral decision-making, which indicates ways of using other behavioural fields to forge the moral character of organisations more effectively. However, as will be seen over the course of this chapter, the decisions and normative orientation provided for companies will depend on decisions being taken not simply from a biological-brain perspective, but also from the perspective of participative deliberation devoted to what is just, prudent and responsible. The importance of the latter premise must be underscored from the outset and it is explored in the next section.

9.2 What Does It Mean to Forge Moral Character in Organisations?

Organisational ethics implies two levels of thought: descriptive and normative. On the one hand it is important to uncover the values, norms and principles that guide decision-making and are present in organisations' formal management documents. On the other hand, and in addition to a factual description of an organisation, it must also involve a critical, universalist attitude that allows an organisation's conduct and values to be both critiqued and guided. Founded on this concept of integrative ethics (Ulrich 2008) the task of moral philosophy is to work alongside the main actors in business practice—citizens, experts in business activity, and governments (Cortina et al. 2008)—in order to pursue its goals in the practical sphere of human activity and its institutions. However, to maintain the universality of this approach a normative measure must be applied that allows us to rupture the rigidity of the factual, and this is provided by intersubjectivity (Habermas 1998, 1993, 1972), which is achieved through dialogue-based deliberation involving all the stakeholders within an organisation (González-Esteban 2002).

In the pages that follow I argue for the need to identify and create spaces within organisations for the development of scientific, epistemological and practical capabilities, or in other words, cognitive and affiliative skills, the ability to relate to other people and the environment, and also a capacity for dialogue. Thereby, I underscore the importance of organisations creating spaces for deliberation, forging and developing moral character, and fostering knowledge, as well as interest, reasons and emotions. All these aspects have been developed in business ethics since it emerged in the 1970s by drawing on Aristotelian (Solomon 1992), Kantian (Bowie 1999), contractarian (Donaldson and Dunfee 1999) and Rawlsian (Phillips 2003) perspectives, as well as that of discourse ethics (Cortina et al. 1994, 2008), among other branches of philosophy.

According to authors such as Apel and Cortina, ethics as moral philosophy performs three functions: firstly, to define the moral sphere; secondly, to specify its philosophical grounding or justification; and thirdly, to apply the results of the first two functions in different aspects of social life. Therefore, applied ethics is mainly concerned with applying ethical-theoretical principles, unveiled by or drawn from the process of grounding moral principles within different spheres of human praxis. Yet its task is not to apply them directly—at least that is my proposal here. Rather, applied ethics should aid reflection when taking decisions on specific cases. In this sense, *ethics in organisations as applied ethics*, as a sub-discipline, does not offer immediate set responses to set problems, which would clearly be of little help to professionals and organisations.

Applied ethics is a field with its own tools and philosophical-ethical methods, and it uses them to interpret the moment of unconditionality that pertains to the moral fact, which is the case for organisations as dealt with here, yet it is also confronted in the study of other practical sciences such as economics, new technologies, education, the mass media, and so on. As a result, applied ethics, including organisational ethics, has two basic features. The first is its *interdisciplinary nature*, which requires discussion and complementarity with other academic disciplines. With regard to organisational ethics, for instance, this embraces all the disciplines that pursue a knowledge of how human beings relate in order to achieve common goals, and this includes, at the very least, ethical theory, economic theory, organisational theory, psychology, and organisational and industrial sociology. The second feature is the *specific function* of organisational ethics: the interpretation of the *unconditionality* found in each practical sphere, or in other words, the specification of the *principles and norms* that can be *required* for each situation. The latter is essentially concerned with what makes us human, with the questions of justice and solidarity regarding what individuals want for themselves and for others. In short, an ethical perspective allows us to differentiate between what is right or just from what is not, to recognise through deliberation the principles, values or norms that are deemed to humanise us and that we are not willing to renounce. Autonomy, freedom, equality and respect for dignity are just some examples of the principles and values that can be found in any type of organisation (Cortina et al. 1994; Habermas 1987, 1990, 1993).

With regard to the integrative organisational ethics set out in this chapter, which has its roots in discourse ethics, for an organisation to develop its activity or practice legitimately it must take into account four points of reference. Firstly, *the social goal* that will give the organisation meaning: to meet the human needs of the groups affected by its activity to a high standard. Secondly, the appropriate *mechanisms* to achieve these goals: in the case of a business organisation these include the market, competition, and the pursuit of economic profit and social benefit. Thirdly, the *legal-administrative framework* stipulated by an organisation's constitution and current complementary legislation. And, finally, the *critical moral consciousness* achieved by an organisation in society (Cortina et al. 1994: 38–49). These four points of reference are relevant to both the configuration and development of an ethical organisation culture and ethical decision-making processes because they are embedded in moral values, norms and principles. However, I am aware that some of these factors have been subjected to ethical reflection, others have not.

Training in critical thinking and designing spaces within organisations in order to encourage a critical moral consciousness—the capacity that allows us to differentiate *what is* from *what should be*, or *what happens* from *what should happen*—are two essential ways of enabling a real ethical transformation to occur in organisations (Cortina et al. 2008; Habermas 1990). Yet it should be noted that this training does not guarantee ethical changes since the step from theory to practice is not without its difficulties, but it will be an enabling factor. In this chapter I advocate the promotion of critical moral judgements being made in both decision-making processes and through opening up spaces within organisations in the form of codes of conduct and ethics, organisational ethics committees, social responsibility reports, and ethics hotlines. Other tools could also be included in this list, but these four allow all the affected stakeholders to participate in and deliberate on these spaces, which helps to open up them up still further. In addition, these four tools are interconnected in a way that facilitates consistency in the management of an organisation's principles and moral values, as well as its behaviour (García-Marzá 2017).

It is worth noting here that, according to Cortina, in general, education is the starting point for intelligently educating desire and educating intelligence in a desiderative manner (Cortina 2011). Education enables understanding and the learning of traditions and world-views, but it also offers—or at least should offer—intelligent guidelines according to which one can learn how to value, choose, prefer and imagine courses of action.

It may also be proposed that education is moral education in that it is simply the process by which subjects are offered guidelines capable of generating freedom, in the sense that these guidelines grant subjects the capacity to exercise their own autonomy. Therefore, these guidelines are akin to tools that enable subjects to unveil the irrevocable moral minimums of justice that exist in any field of activity they embark upon. According to Habermas, such moral minimums are not merely the product of tradition; rather, they may be discovered within the essence of individuals as intersubjective beings, who are capable of entering into dialogue, and reaching a consensus with their peers that sustains intersubjectivity. In other words, working from the foundation of transcendental pragmatics, our knowledge of what is right, or fair is understood as always being achieved through the interpellation between subjects and the dialogues that take place between them when they try to discover what is right or correct. Therefore, the minimums of justice are not the result of experience or intuitive knowledge, but instead are discovered through dialogues and discourse when another person or situation challenges us.

Thus, for organisations to forge moral character they must implement a critical education based on dialogue that is capable of offering guidance, but does not take decisions on behalf of subjects; or in the words of Stein et al. (2011) that is capable of "raising" not "designing" decision-making actors in organisations. This chapter proposes that one way of achieving such a critical education is through the use of ethical codes, ethics committees, responsibility reports and ethical hotlines as spaces for deliberation and critical education. However, to "raise" a person in the most effective way it is also necessary to know, among other types of knowledge, about the relevant physiological factors and how they are involved, whether it be positively or negatively.

9.3 What Do the *Neuro* Disciplines Tell Us About Forging Moral Character in Organisations?

Several disciplines have made significant developments in recent years thanks to their engagement with the advances made in neuroscience since the 1990s. With regard to the forging of moral character in organisations, three new disciplines are relevant: neuroethics, organisational neuroscience and neuroeducation, all of which share a concern to understand the neural foundations of their respective object of study: ethics, organisations and education.

An analysis of the proposals and advances made, and the results achieved in these three fields with regard to the present discussion reveals several shared features. Firstly, they are all presented as interdisciplinary areas of knowledge that attempt to transfer the implications of advances and ideas about the brain and the mind, acquired through neuroscientific methods, to a specific practice: learning in the classroom, decision-making processes in organisations, etc. (Ansari et al. 2012; Hardiman et al. 2012; Howard-Jones 2014; Howard-Jones and Fenton 2012; Nouri 2016; Palghat et al. 2017). Secondly, there are intrinsic relationships between reason and emotion within both human cognition, regarding which emotions are extremely relevant in moral decision-making, and effective learning, above all in relation to learners' motivation to move from theoretical to practical learning, or from impersonal to personal scenarios (Ansari 2012; Immordino-Yang and Damasio 2007; Narvaez and Lapsley 2008; Narvaez 2012). Thirdly, experts in neuroethics and neuroeducation have cautioned against drawing hasty conclusions, and not extrapolating from descriptions of neural studies and experiments any precipitated theories or conclusions of a normative nature. Expressions typically found in the titles of research papers and the conclusions of a range of publications in these fields include "caution", "not so fast", or "a bridge too far" (Bruer 1997, 2008), and in the field of neuroeducation this is above all due to the concern that drawing premature conclusions may mislead education practice, and result in it being influenced by "neuromyths", or a tendentious knowledge supported by false neuroscientific advances (Ansari 2012; Bruer 1997; Howard-Jones and Fenton 2012; Pallarés-Dominguez 2013). In the words of Hardiman et al., "these naïve misinterpretations of science have spread throughout the folk psychology of educators in recent years" (2012: 136). Finally, it is understood that neuroscientific advances made in the fields of ethics, education and organisation need to move beyond the laboratories and into practical settings, whether that be classrooms and organisations, or in the form of public policy guidelines. Thereby, they can be applied with a view to improving people's lives or well-being (Cortina 2012; Cropanzano and Becker 2013; Evers 2009; Palghat et al. 2017; Racine et al. 2017).

We therefore need to identify the basic features that define each of these three disciplines in order to reveal what they have to offer, and consider how this may further our knowledge of the neural foundations of decision-making and moral judgement. Thereby, from an organisational ethics perspective, the "gains" and advances made by the branches of neuroscience can be used to guide the forging of moral character in organisations. Moral character is understood as the moral values, norms and principles that define an individual or an organisation that are not innate, but are the result of choice, a choice that can be made either heteronomously or autonomously.

Neuroethics can be defined as the field concerned with the two-way relationship between ethics and neuroscience. From the perspective of the ethics of neuroscience, neuroethics seeks to develop and propose a framework to regulate the conduct of neuroscientific research and to inquire into and guide its applications. Viewed from the perspective of the neuroscience of ethics, neuroethics is concerned with exploring the repercussions and impact advances in neuroscience have upon our consideration of ethics, and how they may modify any consideration of ethics, whereby one of its key objectives is to study the neural bases of moral agency (Cortina 2012; Racine et al. 2017; Roskies 2002). However, as an applied ethics, the task of the neuroscience of ethics is not only to discover but also to guide. It is in order to undertake this dual task that ethicists and neuroscientists are required to work side by side exploring research horizons that should, at least in my view, include the following: (a) the cerebral foundations of moral conduct; (b) whether these foundations provide the *grounds* upon which moral obligations are drawn, or are simply a cerebral base, and therefore not a basis for moral life; and (c) the promotion of an interdisciplinary approach that ensures neuroscientific findings are used to benefit society, for example by improving public policies, rather than serving specific economic powers through neuromarketing. Yet this interdisciplinarity is not without difficulties (Castelli 2018).

Since its emergence, neuroethics, with its basis in neuroscientific advances, has shown that moral judgement has emotional as well as cognitive elements (Damasio 1999, 2005, 2010; Dedeke 2015). Emotions are revealed as key variables, along with others such as a subject's distance from a situation, or whether a moral judgement to be made is "personal or impersonal" (Greene et al. 2001).

For its part, neuroeducation, also known as "educational neuroscience" or "mind, brain and education", could be defined as "a nascent discipline that seeks to blend the collective fields of neuroscience, psychology, cognitive science, and education to create a better understanding of how we learn and how this information can be used to create more effective teaching methods, curricula, and educational policy" (Carew and Susan 2010: 685). One of the objectives of neuroeducation is to use neurobiological knowledge to inform the intricate relationships between emotion and cognition in order to address the role of emotion in education more effectively, and thereby help reshape educational research, practice and policy (Battro et al. 2008; Immordino-Yang and Damasio 2007; Stein et al. 2011). In general, studies and advances in neuroeducation focus on practical learning situations with young people, but the proposal made in this chapter is that the progress made by neuroeducation—and within the field of organisational neuroscience— need to be explored in order to guide the training of moral character within organisations.

The process of moral character formation implies, firstly, recognising the values, norms and moral principles that orient specific organisational relations and decision-making processes. Secondly, spaces must be created for establishing a dialogue about any moral conflicts that arise when making decisions or recommendations, as well as creating policies that are better aligned with an organisation's values, norms and principles. It is this ethical culture that is reflected in ethical and behavioural codes. Therefore, each decision-making process along with the organisation's ethical management tools provide spaces for encouraging reflection on both the values held by an organisation and the values it deems to be especially valuable. The processes of the recognition of and dialogue on value conflicts must be seen in their own right as part of the process of forming the organisation's specific moral character. The key therefore lies in using neuroeducational contributions to develop organisations' moral learning processes as a means to improve moral learning across the organisation.

The approach taken in this chapter is based on the view set out by Stein et al. that the purpose of education is to transmit cultural values and skills, whereby the task involves being able to jointly construct, together with the learner, shared objectives and values, rather than inculcating these values (Stein et al. 2011). Therefore, in order to successfully train moral character, it is crucially important that the methods, situations and attitudes employed are framed by first and second person perspectives, and that language and cooperation is foregrounded. The training process, considered from this standpoint, is thus conducted through approaches that lead to dialogue and consensus. In the field of educational neuroscience, psychopharmaceuticals have sometimes been suggested as a way to bring about improvements (including moral improvements), but this direction does not respect the subject's autonomy, since all relationships should be grounded on the structure of mutually reciprocal dialogue (Cortina 2011; Narvaez 2012; Stein et al. 2011). In regard to this advances in neuroeducation have shown that if first-person and second person perspectives are maintained when constructing and acquiring knowledge, then learning is much more effective and long-lasting, and even more so when the brain's plasticity is considered as this means we can acquire new knowledge throughout our lives, regardless of age (Ansari 2012; Narvaez 2012).

Authors like Lieberman and Narvaez highlight the importance of the use of narrative and communication between peers as elements of learning processes (Lieberman 2012; Narvaez 2011). It is thanks to our social brains that we learn more through the support of mental circuits than though memory (Lieberman 2012: 7). Social motivation, life stories and narrative are therefore a necessary part of learning processes. Narvaez, in particular, has developed an educational model known as integrative ethical education (IEE), which is centred on a system of activities based on learner-expert learning, and it applies positive social influences to the brain's connections and behaviour in order to lead to personal and group empowerment (Maxwell and Narvaez 2013; Narvaez and Lapsley 2008; Narvaez and Vaydich 2008; Narvaez 2006, 2007). According to this model deliberation is fundamental for the four-stage process of moral behaviour: moral sensitivity, moral judgement, moral motivation and moral action (Rest 1986; Rest et al. 2000). The IEE model explains how the moral character of individuals and organisations develops and the elements involved in this development, and it also provides neurobiological and neuroscientific evidence for each of the four stages of moral behaviour.

The third "neuro" field related to organisations is organisational neuroscience. It was first mooted in 2007 by Michael Butler and Carl Senior as a new research field capable of generating new knowledge, one that would go beyond the separate discourses of cognitive neuroscience and organisational theory. Butler and Senior argue that the great debate concealed within organisational neuroscience is the role played by nature on one side, and culture and education on the other, both with regard to individuals and organisations. Organisational cognitive neuroscience uses tools like neuroimaging to analyse the human brain and understand its cerebral function, and in turn human behaviour within an organisational framework (Butler and Senior 2007: 8). To date, this line of research has produced various organisational neuroscience models, which are based on three main assumptions (Becker et al. 2011; Becker and Cropanzano 2010; Cropanzano and Becker 2013).

The first of these assumptions claims that the workings of mirror neurons underpin many group-adaptive behaviours. The explanation given by organisational neuroscience for the theory of mirror neurons is that our brains are connected in such a way that we are open to social influences to ensure we can adapt, whereby mirror neurons provide a physiological basis to explain human behaviour. Therefore, we have a physiological basis that predisposes us to vicarious learning, and this can be made use of at an organisational level. Furthermore, this knowledge extends to what we know about how people interact in generating organisational cultures, and individuals' attitudes to teamwork. When a group interacts, it learns from the value judgements and behaviour of the other group members. On occasions this may give rise to attitudes or behaviour that is adaptive for the group, but does not necessarily achieve an objective that legitimises the organisation, whether this is health, education, the production of a good or service, etc.; such behaviour may include corruption, nepotism or malpractice. Feelings and emotions are crucial to this vicarious learning, since they allow unconscious learning to occur very swiftly among members of the same group (Becker et al. 2011; Becker and Cropanzano 2010; Cropanzano and Becker 2013).

Secondly, neuroscientific and neuropsychological findings indicate that our brains unconsciously process much of what we learn during socialisation, so we are often unaware of the implicit attitudes underlying our behaviour. These implicit attitudes appear automatically and very suddenly in our brain, and it is only when we experience a moment of confusion or confrontation that we begin to rationalise them (Haidt 2001). Neuroscience confirms the physiologically based neural explanation that demonstrates the difference between explicit and implicit attitudes. Regarding explicit attitudes, studies highlight the role of deliberative reasoning that arises in the executive controlling structures of the frontal lobe, which are relatively new in evolutionary terms, whereas implicit attitudes are produced phylogenetically in the brain's older regions. In an organisational context this could explain why cultural or ethical organisational changes are difficult to implement because companies and their managers focus on explicit attitudes and fail to appreciate and take into account the role played by implicit attitudes. Sometimes not even employees themselves are aware of their own reticence to cultural or ethical change, since it is implicit attitudes that arise first and are able to prevent other implicit attitudes or conscious processes from emerging. It is this factor that may also explain employment discrimination in the form of intolerance in multicultural environments, or it may even be the result of a disconnection between implicit and explicit attitudes. In my view, these cognitive dissociations require the development and implementation of organisational ethical policies that can serve as explicit guidelines for decision-making processes based on shared values. They therefore allow an organisation's moral character to be established, both in the professional field and on an individual basis.

Finally, organisational neuroscience has arisen as a new area of knowledge that can enhance our understanding of numerous aspects of business relationships, as well as the motivations for action that underlie the organisation's values. For example, with regard to the issue of justice in labour environments Folger and Salvador identify two major positions within the organisational literature (Folger and Salvador 2008). On one hand, some authors contend that individuals often value fairness for its own sake, that is, regardless of the outcome it has for themselves or for the organisation; it is considered as a value in itself. In contrast, other authors hold that justice is not a value in itself, but a means by which the individual can gain an advantage, and thus individuals are only motivated to defend justice when their own interest is at stake. The debate on self-interest is obviously complex and involves many disciplines, including philosophy, psychology, and business management and administration. However, neuroscientific advances and experiments suggest that we are hard-wired to cooperate, and that we have a phylogenetic innate sense of justice that precedes later cognitivist rationalisation. There is therefore a need to revisit the naïve vision of rational economic agents as the maximiser of their own usefulness, which leaves selfishness as the only possible guide for behaviour and decision-making (Conill 2017). But in my view, the most relevant insight provided by the discipline of organisational ethics is that neuroscientific studies suggest we have a moral sensitivity towards justice that is related to our emotional responses. And it is these emotional responses that can be educated, or put another way, oriented and guided, because culture is considered to have a major influence on the way this specific sense of justice is developed.

In sum, these three new fields shed light on cerebral activity in situations of cooperation, learning processes, educational modelling, and in processes of moral decision-making and reaching moral judgements. However, the "framework" they provide is not the same as a "grounding", and the difference depends on our having reasons to act with justice, prudence and responsibility in organisations.

In accordance with the discourse ethics proposal made in the 1980s by K. O. Apel (1998) and J. Habermas (1990, 1993), the foundation of moral obligation lies in norms, which are none other than reciprocal expectations of behaviour generalised through the norms with which we organise our lives in a coordinated way (Habermas

1990, 1996). Cortina has stated that five traits are required for this initial position of the grounding for moral obligation, set out within discourse ethics: (a) social life is grounded on subjects' mutual recognition of each person's communicative competence; (b) based on this recognition, links are established that generate obligations between subjects and that go beyond kinship or proximity; (c) obligations are not only established by direct or indirect reciprocity; (d) recognition goes beyond the calculation of the largest number to refer to all human beings; and (e) obligation does not rest only on survival or prudence, but on the right of all beings with communicative competence to be treated as valid interlocutors (Cortina 2016).

It is the philosopher's task to provide ethical frameworks that can be applied through an enquiry into the grounding of a moral obligation and the clarification of why it is understood as "moral"; in this chapter moral obligation is grounded in intersubjectivity. Similarly, the cerebral foundations of behaviour and the moral progress that has culturally impregnated our moral way of knowing must also be taken into account (Bowie 2009; Cortina 2012). Therefore the role of experts in organisational ethics is to use what has been learned and gained from the three *neuro* disciplines discussed above, in order to explore organisational practice, as well as create and guide the design of spaces for creating moral judgement in organisations, while also working towards developing them through practice.

9.4 How to Design Spaces for Creating the Optimum Moral Judgements Within Organisations

As business ethics has shown over the last 25 years or more, forging—or building—moral character in organisations involves influencing processes and organisational values, in other words, business policies, as well as strategy and organisational culture (Cortina et al. 1994; González-Esteban 2012). This entails developing an ethical conscience within an organisation, as well as an organisational decision-making structure that recognises the ethical dimension of decision-making processes. Education or training in organisations is therefore essential, since it is not sufficient to merely design procedures without guiding individuals' character, which is the original sense of education.

Consistent with the aforementioned proposal of an organisational discourse ethics, a set of tools can be used to manage trust responsibly by integrating transparency and participation within an organisation. In particular, this refers to the design of procedures that are integrated into a system for managing ethics within organisations; this comprises a code of ethics and conduct, the creation of an ethics committee, a social responsibility or sustainability report, and, finally, an ethics hotline (García- Marzá 2017). These tools are proposed as an ethics programme because they involve a level of reflexivity and enable the creation of spaces for both formal and informal deliberation in an organisation. Needless to say, there are other tools that facilitate the development of ethics practices, but not all of them fulfil the requirements of reflexivity and deliberative intersubjectivity.

The *code of ethics and conduct* is a formal document that explicitly describes the values and commitments an organisation pledges to implement. It is drawn up with the participation of all the organisation's stakeholders and has internal and external implications. With regard to its internal implications, it is the expression of the organisation's character and its way of being, and acts as a reference for decision-making and the building of the organisation's specific culture. Its external implications relate to the means by which an organisation presents itself to society, as well as its global relationships (Lozano-Aguilar 2016; Schwartz 2004). The preparation of the document and its subsequent follow-up are in themselves spaces for participation and deliberation within organisations.

The *ethics committee* is understood as a space for participation and deliberation where all internal and external stakeholders are represented. It has three functions: (a) to prepare reports on suggestions, alerts and complaints about the code of ethics and conduct, or its application, which have been raised via the ethics hotline; (b) to advise on the interpretation of the code of ethics and conduct and its application; and (c) to promote training for employees and managers on the code of ethics and conduct, and on use of the ethics and compliance hotline (García-Marzá 2017).

The *responsibility or sustainability report* is usually published annually and provides information on the organisation's impact upon its economic, social and environmental context. It therefore addresses specific actions that have been adopted during a given period on matters and issues of interest to its stakeholders. It also reflects upon the way the organisation has acted in accordance with its code of ethics and conduct. The various types of impact described in such reports, as well as being verifiable must also be related to the organisation's principles, values and commitments (Sethi et al. 2017). Reports of this type have become increasingly standardised and they not only provide information on the organisation's policies, strategies and culture, but also on how it has achieved its accomplishments, for which quantitative and qualitative data on indicators is used, which can be compared and contrasted on a yearly basis.

The *ethics hotline* can be used to report activities that contravene the organisation's code of ethics and conduct, to warn of dubious situations or future risks, and to make suggestions for improvement (Calvo 2018). Its main aim is to establish an open space for participation where responsibility for ethical action is shared and stakeholders feel they have equal responsibility for its development.

Together these four spaces for participation and deliberation are aimed at "creating a culture where transparency and participation discourage bad practices and acknowledge and strengthen good practices" (García-Marzá 2017: 273). As I have argued throughout this paper, in building such a culture there must be spaces where an organisation's moral judgement can be formed and that the design and use of these spaces should take into account advances from neuroethics, neuroeducation and organisational neuroscience.

In particular, five aspects need to be highlighted: (1) a recognition of the importance of vicarious learning; (2) the existence of a moral sensitivity to justice that is related to emotional responses; (3) implicit attitudes that the brain has processed unconsciously, which are supported through non-reflective social prototypes; (4) the importance of first and second person narratives, as well as experience in learning processes; and (5) a complex concept of morality that integrates intuition and reasoning.

Some studies have, on occasions, hastily reported neuroscientific findings in order to explain moral behaviour, or else they have presented them as a naturalisation of morality. However, since the turn of the century, in the fields of psychology (Narvaez), neuroethics (Evers and Salles), neuroeducation (Stein), organisational neuroscience (Becker and Cropanzano) and philosophy (Cortina), such positions or studies have been questioned on the grounds that the contribution of neuroscientific advances in these fields solely sheds light on the neural and cerebral bases of human moral behaviour, but not on its grounding or normativity. The latter function should, I believe, be carried out from a universalist perspective of critical and post-conventional ethics.

Drawing on the knowledge that neuroscience has provided of the cerebral foundations of human behaviour, I want to now propose a series of guidelines for training moral judgement with in the aforementioned spaces for participation and dialogue, and also discuss how these engage with advances made by the aforementioned *neuro* disciplines.

9.5 Guidelines for Moral Training in Organisations

First and foremost, a succinct review is required of some of the lines of enquiry that have been developed with regard to organisations' spaces for participation and deliberation, and in particular how these involve managing the moral character required by an organisation. In my view, it is necessary to include a series of guidelines for forging moral character within organisations, guidelines that foreground the role of emotions, reason, affectivity and cognition simultaneously. In addition to an integrated ethics management model that includes spaces for participation, deliberation and ethics training, there is also a need for guidelines on how to shape ethical character within organisations more effectively in order to help organisations make the best moral judgements (González-Esteban 2016).

Below I set out six areas within which work can be done to forge—or shape—the moral character of organisations, and where neuroscience, drawing on recent research from organisational neuroscience, neuroeducation and neuroethics, indicates the directions that may prove to be most effective and influential. It is not my intention to propose a full model for forging moral character, but as a preliminary stage I set out a series of general guidelines for action that can help guide the appropriate training within organisations. Both the strengths and the weaknesses of the guidelines are presented in what follows.

Guideline 1 Consider both cognitive (interest, norms, values) and affective (emotions and feelings) aspects of ethical decision-making processes. However, as Bowie (2009) states, showing how decision-making processes work is not the same as justifying a decision, so the development of decision-making must take into account the accompanying complex web of emotions, intuitions and reasons, yet without using the latter to morally justify the decisions taken (Lozano-Aguilar 2016). Normative-philosophical developments such as the discourse-organisational ethics proposal should offer ways of further developing this function.

Guideline 2 Encourage the unveiling and definition of good practices and their establishment as non-reflective social prototypes for training in organisations. As neuroethics research has shown, in making ethical decisions, emotions—and perhaps also the non-reflective and unconscious processes linked to prototypes—all play a more predominant role in day-to-day activity than ethical rules or ethical reflectivity. Therefore, organisational ethics must develop guidelines for good practice that can be applied to each organisational area and sector, and that can also be used to form a global organisational moral character. Examples would include good practice in human resource management, or supply chain ethics.

Guideline 3 Endeavour to explicitly articulate the moral intuitions in stakeholder relationships so they can be analysed and strengthened or challenged in such a way that it is the organisation that steers the direction of emotions and not the other way round, thereby emotions will be governed by the recognition gained from intuitions. Or in other words, an organisation's moral intuitions must be identified and explicitly acknowledged, and the creation of intersubjective spaces can help this to occur, whereby the emotional response an organisation seeks to give becomes consciously modulated. In line with this, empirical developments and normative proposals must work side by side in organisational ethics. In order to address stereotypes, I would suggest the use of training sessions, for example, that allow each stakeholder group to reflect on the values set out in the code of ethics, or else on conflicts that may be reported to the ethics hotline or raised by the ethics committee. These sessions should use a deliberative methodology that creates a climate in which it is possible to configure moral judgements that reveal both their explicit (reasonings) and implicit (emotions and tacit assumptions) components (Narvaez and Lapsley 2008; Narvaez 2010).

Guideline 4 Study interests from the perspectives of both explicit reasons and the implicit intuitions. The management of relationships between an organisation's stakeholders—those affected by the achievement of the organisation's objectives (Freeman et al. 2010; Freeman 1984; González-Esteban 2007)—should not only focus attention on making explicit and discussing the reasons for their interests, but also on specifying the intuitions at work behind these relationships and emotions. This task clearly calls for organisational processes designed to favour dialogue-based on "reciprocal recognition in the organisation." That is, all stakeholders must be considered as valid interlocutors in the processes of deliberation or discussion of the interests that affect them, so they can explicitly defend their interests in the first-person, and not depend on the interpretation of intermediaries or authorised voices. In this vein, using participatory methods to draw up codes of ethics and conduct that can be reviewed every 2 or 3 years, or setting up ethics committees and ethics and compliance hotlines, spaces for participation and moral judgement can be opened up and used to provide training for new situations or on issues about which moral consensus has yet to be reached.

Guideline 5 Promote dialogue-based, deliberative ethics training in organisations. The content learned is just as important as the way learning occurs. Content must be based on dialogue and narrative so as to prioritise education rather than instruction. Neurocognitive models suggest that ethical education is possible, but in order to be effective, the specific cycle being worked on must be taken into account. In this vein, ethical education must focus on the mental structures of prototypes and moral rules, together with higher order conscious reasoning, which is both based on post-decision rationalisation and deliberative. One of the main functions of this higher order conscious reasoning cycle is to restructure and refine an organisation's prototypes so that they reflect, as appropriately as possible, the prototypes shared by the relevant sociocultural units that are, wherever possible, deliberately constructed (Reynolds 2006). In other words, training the moral character within an organisation must affect the whole process of reasoning and the formation of moral judgements (Dedeke 2015; Narvaez and Lapsley 2008; Schwartz 2004): recognition of the moral situation, moral judgement (reason-emotion), moral reflection and moral intention. Moral deliberation can be understood from this perspective as a constant to and fro between reasoning and intuition, between principles and objectives, between the recognition of assumptions, before coming to a dual reflection, that is both internal, and undertaken as a group. Over the course of this process disputed values, principles or norms can be subjected to scrutiny.

Guideline 6 Identify shared prototypes in the organisation in order to strengthen or reorient them normatively in the event that they clash with the ethical normativity that has been revealed through dialogue. This guideline has implications for organisational culture both in terms of the recognition of the corporate culture itself, and for mechanisms used to modify it and therefore forge moral character (Reynolds 2006). Organisational culture consists of formal and informal structures. The formal structures include the organisation's policy and strategy, which are reflected, for instance, in its systems of compensation and productivity assessment. The informal mechanisms may be reflected in the behaviour of its leaders, or in external pressure. Corporate cultures sometimes differ from wider shared social values, norms and principles, and may be unable to take up critical positions vis-à- vis those that society accepts as valid but are actually only assumed in an acritical manner. According to Reynolds (2006) this can occur for four reasons. Firstly, because the prototypes shared across an organisation are badly labelled or identified, which leads to disharmony with society or among employees. Secondly, because these prototypes assign incorrect normative values to particular actions, which gives rise to responses that infringe upon socially shared values, or those that are relevant to a professional deontology. Thirdly, because an organisational culture emphasises a moral rule that is inappropriate or gives rise to vague prescriptions. And finally, because the moral rules of an organisation are not applied appropriately. Any one of these conditions can lead to unethical behaviour. Therefore, the mechanisms for steering organisational culture towards a more ethical direction lie in understanding the prototypes that are being shared by the organisation as well as its moral rules because, either formally or informally, they are being culturally reinforced. Once these prototypes have been revealed, they can be re-examined and replaced by others that result in an unequivocally ethical behaviour.

9.6 Conclusion

Interdisciplinary approaches and the collaboration between neuroscience and normative science have become increasingly necessary to deal with and guide complex social practices such as those found in organisations. This chapter has proposed a dialogue between intersubjective critical organisational ethics and the advances in *neuro* knowledge made in the fields of ethics, education and organisations.

It has set out to identify the spaces within organisations where their moral judgement can be developed, and it has also attempted to guide the way that this can be achieved in the light of recent research in the fields of neuroethics, neuroeducation and organisational neuroscience. Organisational ethics aims to guide the organisation's activities along lines of justice, prudence and responsibility. For this purpose, it is essential to open up spaces within organisations where this dialogue becomes possible and that can favour the configuration and training of post-conventional critical moral judgements. What this chapter has sought to show is how neuroscientific reflection can shed light on how organisational learning can be both improved and applied in four specific spaces—codes of ethics and conduct, social responsibility or sustainability reports, ethics committees and ethics hotlines—through the use of deliberative methodologies.

References

Ansari, Daniel. 2012. Culture and education: New frontiers in brain plasticity. *Trends in Cognitive Science* 16 (2): 93–95.

Ansari, Daniel, Bert De Smedt, and Roland H. Grabner. 2012. Neuroeducation – A critical overview of an emerging field. *Neuroethics* 5 (2): 105–117.

Apel, Karl-Otto. 1998. *Towards a transformation of philosophy*. Milwaukee: Marquette University Press.

Battro, Antonio M., Kurt W. Fischer, and Pierre J. Léna. 2008. Introduction: Mind, brain, and education in theory and practice. In *Educated brain: Essays in neuroeducation*, ed. Antonio Battro, Kurt W. Fischer, and Pierre J. Léna, 3–19. Cambridge: Cambridge University Press.

Becker, William J., and Russell Cropanzano. 2010. Organizational neuroscience: The promise and prospects of an emerging discipline. *Journal of Organizational Behavior* 31: 1055–1059.

Becker, William J., Russell Cropanzano, and Alan G. Sanfey. 2011. Organizational neuroscience: Taking organizational theory inside the neural black box. *Journal of Management* 37 (4): 933–961.

Bowie, Norman E. 1999. *Business ethics. A Kantian perspective*. Oxford: Blackwell Publishers.

———. 2009. How empirical research in human cognition does and does not affect philosophical ethics. *Journal of Business Ethics* 88: 635–643.

Bruer, John T. 1997. Education and the brain: A bridge too far. *Educational Researcher* 26: 4–16.

———. 2008. Building bridges in neuroeducation. In *Educated brain: Essays in Neuroeducation*, ed. Antonio Battro, Kurt W. Fischer, and Pierre J. Léna, 43–58. Cambridge: Cambridge University Press.

Butler, Michael, and Carl Senior. 2007. Toward an organizational cognitive neuroscience. *Annals of the New York Academy of Sciences* 1118: 1–17.

Calvo, Patrici. 2018. *The cordial economy – Ethics, recognition and reciprocity*. Cham: Springer.

Carew, Thomas J., and H. Susan. 2010. Magsamen neuroscience and education: An ideal partnership for producing evidence-based solutions to guide 21st century learning. *Neuron* 67 (9): 685–688.

Castelli, Paula. 2018. ¿Qué modelo de interdisciplinariedad requiere la neuroética? *Recerca. Revista de Pensament I Anàlisi* 22: 33–49.

Conill, Jesús. 2017. Nuevos horizontes de economía ética en tiempo de neurociencia. *Pensamiento* 73 (276): 541–562.

Cortina, Adela. 2011. *Neuroética y neuropolítica. Sugerencias para la educación moral*. Madrid: Tecnos.

———. 2012. Neuroética: presente y futuro. In *Guía Comares de Neurofilosofía práctica*, ed. Adela Cortina, 9–38. Granada: Comares.

———. 2013. Corporate social responsibility and business ethics. In *Corporate citizenship, contractarianism and ethical theory. On philosophical foundations of business ethics*, ed. Jesús Conill, Christoph Luetge, and Tatjana Schönwälder- Kuntze, 69–78. Aldershot: Ashgate Publishing.

———. 2016. La conciencia moral desde una perspectiva neuroética. De Darwin a Kant. *Pensamiento* 72 (273): 771–788.

Cortina, Adela, Jesús Conill, Agustín Domingo, and Domingo García-Marzá. 1994. *Ética de la empresa. Hacia una nueva cultura empresarial*. Madrid: Trotta.

Cortina, Adela, Domingo García-Marzá, and Jesús Conill, eds. 2008. *Public reason and applied ethics. The ways of practical reason in a pluralist society*. Aldershot: Ashgate Publishing.

Cropanzano, Russell, and William J. Becker. 2013. The promise and peril of organizational neuroscience: Today and tomorrow. *Journal of Management Inquiry* 22 (3): 306–310.

Cropanzano, Russell, Sebastiano Massaro, and William J. Becker. 2017. Deontic justice and organizational neuroscience. *Journal of Business Ethics* 144: 733–754.

Damasio, Antonio. 1999. *The feeling of what happens: Body and emotion in the making of consciousness*. New York: Harcourt Brace.

———. 2005. *Descartes' error: Emotion, reason and the human brain*. Rev. ed. London: Penguin.

———. 2010. *Self comes to mind: Constructing the conscious brain*. New York: Pantheon.

Dedeke, Adenekan. 2015. A cognitive-intuitionist model of moral judgment. *Journal of Business Ethics* 126: 437–457.

Donaldson, Thomas, and Tomas W. Dunfee. 1999. *Ties that bind. A social contract approach to business ethics*. Boston: Harvard Business School Press.

Evers, Kathinka. 2009. *Neuroética. Cuando la materia se despierta*. Madrid: Katz.

Farah, Martha. 2002. Emerging ethical issues in neuroscience. *Nature Neuroscience* 5: 1123–1129.

Folger, Robert, and Rommel Salvador. 2008. Is management theory too "self-ish"? *Journal of Management* 34 (6): 1127–1151.

Freeman, Richard E. 1984. *Strategic management. A stakeholder approach*. Toronto: Pitman.

Freeman, Richard E., Jeffrey S. Harrison, Andrew C. Wicks, Bidhan L. Parmar, and Simone de Colle. 2010. *Stakeholder theory. The state of the art*. Cambridge: Cambridge University Press.
García-Marzá, Domingo. 2017. From ethical codes to ethical auditing: An ethical infrastructure for social responsibility communication. *El profesional de la información* 26 (2): 268–276.
García-Marzá, Domingo, and Ramón Feenstra. 2013. *Ética y neurociencias. La aportación a la política, la economía y la educación*. Castelló: Servei de Publicacions Universitat Jaume I Castelló.
González-Esteban, Elsa. 2002. Defining a post-conventional corporate moral responsibility. *Journal of Business Ethics* 39 (1–2): 101–108.
———. 2007. La teoría de los stakeholders: un puente para el desarrollo práctico de la ética empresarial y de la responsabilidad social corporative. *Veritas. Revista de Filosofía y Teología* 2 (17): 205–224.
———. 2012. Formación ética de los profesionales. Forjando el interés desde la razón y la emoción. *Revista Internacional de Organizaciones* 10: 21–40.
———. 2016. El desarrollo de la ética empresarial ante los avances de la neurociencia organizacional y la neuroética. *Pensamiento* 72 (273): 921–940.
Green, Joshua D., R. Brian Sommerville, Leigh R. Nyström, John M. Darley, and Jonathan D. Cohen. 2001. An fMRI investigation of emotional engagement in moral judgment. *Science* 293: 2105–2108.
Habermas, Jürgen. 1972. *Knowledge and human interests*. Boston: Beacon Press.
———. 1987. *The theory of communicative action*. Vol. 2. Boston: Beacon Press.
———. 1990. *Moral consciousness and communicative action*. Cambridge: The MIT Press.
———. 1993. *Justification and application: Remarks on discourse ethics*. Cambridge: The MIT Press.
———. 1996. *Between facts and norms. Contributions to a discourse theory of law and democracy*. Cambridge: Polity Press.
———. 1998. The inclusion of the other. In *Studies in political theory*. Cambridge: The MIT Press.
Haidt, Jonathan. 2001. The emotional dog and its rational tail: A social intuitionist approach to moral judgment. *Psychological Review* 108 (4): 814–834.
Hardiman, Mariale, Luke Rinne, Emma Gregory, and Julia Yarmolinskaya. 2012. Neuroethics, neuroeducation, and classroom teaching: Where the brain sciences meet pedagogy. *Neuroethics* 5 (2): 135–143.
Howard-Jones, Paul A. 2014. Neuroscience and education: Myths and messages. *Nature Reviews. Neuroscience* 15 (12): 817–824.
Howard-Jones, Paul A., and Kate D. Fenton. 2012. The need for interdisciplinary dialogue in developing ethical approaches to neuroeducational research. *Neuroethics* 5 (2): 119–134.
Illes, Judy, ed. 2006. *Neuroethics: Defining the issues in theory, practice, policy*. Oxford: Oxford University Press.
Immordino-Yang, Mary H., and Antonio Damasio. 2007. We feel, therefore we learn: The relevance of affective and social neuroscience to education. *Mind, Brain, and Education* 1 (1): 3–10.
Levy, Neil. 2007. *Neuroethics*. New York: Cambridge University Press.
Lieberman, Matthew D. 2012. Education and the social brain. *Trends in Neuroscience and Education* 1: 3–9.
Lindebaum, Dirk, and Mike Zundel. 2013. Not quite a revolution: Scrutinizing organizational neuroscience in leadership studies. *Human Relations* 66 (6): 857–877.
———. 2016. Transformación neuroética de la racionalidad económica. Implicaciones para tomar decisions en las empresas. *Pensamiento* 72 (273): 901–919.
Maxwell, Bruce, and Darcia Narvaez. 2013. Moral foundations theory and moral development and education. *Journal of Moral Education* 42 (1): 271–280.
Narvaez, Darcia. 2006. Integrative ethical education. In *Handbook of moral development*, ed. Melanie Killen and Judith G. Smetana, 703–732. Mahwah: Erlbaum.

———. 2007. Triune ethics: The neurobiological roots of our multiple moralities. *New Ideas in Psychology* 26: 95–119.

———. 2010. Moral complexity: The fatal attraction of truthiness and the importance of mature moral functioning. *Perspectives on Psychological Science* 5 (2): 163–181.

———. 2011. The ethics of neurobiological narratives. *Poetics Today* 32 (1): 81–106.

———. 2012. Moral Neuroeducation from early life through the lifespan. *Neuroethics* 5 (2): 145–157.

Narvaez, Darcia, and Daniel K. Lapsley. 2008. Teaching moral character: Two strategies for teacher education. *The Teacher Educator* 43 (2): 156–172.

Narvaez, Darcia, and Jenny Vaydich. 2008. Moral development and behaviour under the spotlight of the neurobiological sciences. *Journal of Moral Education* 37 (3): 289–313.

Nouri, Ali. 2016. The basic principles of research in neuroeducation studies. *International Journal of Cognitive Research in Science, Engineering and Education* 4 (1): 59–66.

Palghat, Kelsey, Jared C. Horvath, and Jason M. Lodge. 2017. The hard problem of 'educational neuroscience'. *Trends in Neuroscience and Education* 6: 204–210.

Pallarés-Dominguez, Daniel. 2013. Críticas y orientaciones para el estudio en neuroética. *Recerca. Revista de Pensament i Análisi* 13: 85–102.

Phillips, Robert A. 2003. *Stakeholder theory and organizational ethics*. San Francisco: Berrett-Koehler.

Racine, Eric, Veljko Dubljevic, Ralf J. Jox, Bernard Baertschi, Julia F. Christensen, Michele Farisco, Fabrice Jotterand, Guy Kahane, and Sabine Müller. 2017. Can neuroscience contribute to practical ethics? A critical review and discussion of the methodological and translational challenges of the neuroscience of ethics. *Bioethics* 31 (5): 328–337.

Rest, James R. 1986. *Moral development: Advances in research and theory*. New York: Praeger.

Rest, James R., Darcia Narvaez, Stephen J. Thoma, and Muriel J. Bebeau. 2000. A Neo-Kohlbergian approach to morality research. *Journal of Moral Education* 29 (4): 381–395.

Reynolds, Scott J. 2006. A neurocognitive model of the ethical decision-making process: Implications for study and practice. *Journal of Applied Psychology* 91 (4): 737–748.

Roskies, Adina. 2002. Neuroethics for the new Millenium. *Neuron* 35: 21–23.

Schwartz, Mark S. 2004. Effective corporate codes of ethics: Perceptions of code users. *Journal of Business Ethics* 5 (4): 321–341.

Sethi, S. Prakash, Terrence F. Martell, and Mert Demir. 2017. Enhancing the role and effectiveness of corporate social responsibility (CSR) reports: The missing element of content verification and integrity assurance. *Journal of Business Ethics* 144: 59–82.

Solomon, Robert C. 1992. *Ethics and excellence. Cooperation and integrity in business*. Oxford: Oxford University Press.

Stein, Zachary, Bruno della Chiesa, Christina Hinton, and Kurt W. Fischer. 2011. Ethical issues in educational neuroscience: Raising children in a brave new world. In *The Oxford handbook of neuroethics*, ed. Judy Illes and Barbara J. Sahakian, 803–823. Oxford: Oxford University Press.

Ulrich, Peter. 2008. *Integrative economic ethics: Foundations of a civilized market economy*. Cambridge: Cambridge University Press.

Chapter 10
Moral Neuroeducation: Proactive Epigenesis and Poverty

Daniel Pallarés-Domínguez

10.1 Introduction: Bridging the Gap Between Neuroethics and Neuroeducation

Concerns about character development are shared by most social educational agents—from parents to philosophers— across all contexts, especially in schools where children spend considerable time away from their family. However, it has to be asked, how effective is character education? And what strategies can be developed for its use? For some time now, emphasis has been placed either on curriculum design, or on learning disorders that arise during the educational process (Sigman et al. 2014). What has been lost sight of is the fact that moral values are immanent to life in school, that is, education has an inescapable moral dimension. However, this is not only with regard to character education in the hidden curriculum, but rather to how and with what methods character educations should be undertaken (Lapsley et al. 2013).

Neuroscience has since begun to study the foundation of morality in relation to decision-making, free will, and the relationship between emotion and reason in moral judgements. Regrettably, some of the results produced so far have proven to be quite puzzling. For example, according to Churchland, the neurological foundation of morality is caused by a higher concentration of oxytocin, vasopressin and norepinephrine, combined with attachment and care for those nearby. Whereas, Haidt's social intuitionism places the emphasis on instincts, and the dual theorist,

This study forms part of the project funded by a postdoctoral research grant [APOSTD/2017/003] from the *Consellería de Educación, Cultura y Deporte de la Generalitat Valenciana* and Fondo Social Europeo. It also contributes to the research project: "Neuroeducación Moral para las Éticas Aplicadas" [FFI2016-76753-C2-2-P] funded by the *Ministerio de Economía y Competitividad.*

D. Pallarés-Domínguez (✉)
Universitat Jaume I, Castellón de la Plana, Spain
e-mail: dpallare@uji.es

P. Calvo, J. Gracia-Calandín (eds.), *Moral Neuroeducation for a Democratic and Pluralistic Society*, https://doi.org/10.1007/978-3-030-22562-9_10

J. Greene, criticises the automatic mode based on emotional responses. Instead, Greene's proposal is to encourage a kind of "deep pragmatism" that leads to a form of rational calculation independent of emotions. However, various authors have argued that this latter model is deficient for many reasons, and a key critique of it is that rationality is understood solely in terms of a utilitarian framework and an emotivist deontologism (Gracia-Calandín 2018). Although none of the above authors focus on the implications of their theories for moral education, their views are likely to have implications. Therefore, how should these supposedly neurobiological and "empirically based" proposals of neuroethics be addressed? The proposal set out in this chapter is that a new field of study devoted to the educational implications of neuroscientific insights on morals needs to be developed. It is time to move from neuroethics to moral neuroeducation, but what does it mean to move from neuroethics to moral neuroeducation? It means that consideration must be given to how the conclusions of both theoretical and practical neuroethics is transmitted and taught. It also means exploring human moral enforceability in relation to the transmission and implementation of knowledge about the brain for the purpose of improving social coexistence (Pallarés-Domínguez and Richart 2018). However, the transmission and implementation of knowledge should not be undertaken using a closed self-sufficient model that derives concrete instructions for behaviour from the interpretation of brain studies. To guarantee the autonomy and freedom of the human being, a reflexive, critical, and dialogical perspective must be employed when addressing neuroscientific proposals (Evers et al. 2017; Cortina 2012; García-Marzá and Feenstra 2013). Or in other words, neuroethical discoveries about human morality cannot be addressed in isolation nor implemented without critical reflection. Therefore, there is a need for a moral neuroeducation that complements neuroethics.

To explore this issue, the discussion pursued in this chapter is divided into three main parts. Firstly, I provide a brief summary of what is understood by neuroeducation. Following this, a critical approach is developed to address the concept and practice of moral neuroeducation. Finally, two examples of poverty are discussed, which reveal how neuroeducation could be developed by employing proactive epigenesis as a core theory of the brain.

10.2 The Basics of Neuroeducation

What is neuroeducation? Is it just another example of the current trend of adding the prefix *neuro-* to fields within the social sciences? Those who remain sceptical in this regard still consider it to be a mistake to link neuroscience with education (Bruer 1997, 2008). However, for others, neuroscience can provide useful knowledge about the brain, and they see neuroeducation as a youthful field of study with great potential. Although, I have defined, conceptualised and classified neuroeducation in other works (Pallarés-Domínguez 2016) a brief conceptual summary is presented here.

Neuroeducation has been understood since its birth in the "decade of the brain" (1990s) as an interdisciplinary field that studies developmental learning processes in the human brain. It was one of the last social neurosciences to emerge and it began to acquire special relevance when the OECD published the report "Understanding the Brain: Towards a New Learning Science" (2002).[1] It encompasses the progressive introduction of neuroscientific studies into learning and teaching methodologies (Goswami 2004). Some scholars have considered neuroeducation mere as an applied cognitive neuroscience (Campbell 2011: 8), while others have sought to define it as a new interdisciplinary science that promotes a better integration and collaboration between the science of learning and the science of neurocognitive development (Koizumi 2008).

Nevertheless, what does this mean? Merely talking about interdisciplinarity or collaboration without defining the ways in which it can be done does not contribute very much. Yet going into more detail could raise additional risks because a relationship between neuroscience and education could lead to the direct application of neuroscientific results in the classroom, and also allow neuroscience to guide educational theory and practice. Such a development would give rise to complaints from many educators would (Howard-Jones 2010), therefore caution is required and both pedagogical and philosophical perspectives must guide collaboration.

In the meantime, neuroeducation has been acquiring normative and methodological relevance due to its results. Taking into account the neural bases of human learning, neuroeducation not only seeks to develop theoretical knowledge, but also addresses practical implications such as the creation of new methodologies, reviewing classical assumptions of learning and studying learning disorders (Tokuhama-Espinosa 2011) (Fig. 10.1).

In a broad sense, neuroscientific techniques have played an important role in developing educational applications. They have been able to offer news solutions, such as the early detection of special learning needs, and even the monitorisation and comparison of different methodologies and individual learning pathways (Goswami 2004: 6). The application of neuroscience to learning processes has also had great impact on the study of early upbringing processes, especially when synaptogenesis is relevantly higher—from 0 to 7 years— (The Royal Society 2011).

Among its various advantages, neuroeducation has also enabled a greater study of other areas related to learning, such as the relationship between sleep and cognition, the importance of physical exercise, learning and time, or chronoeducation, and above all the relationship between emotion and cognition in teaching-learning processes. However, undoubtedly the most important results have been achieved regarding language and mathematics acquisition processes, as well as the teaching

[1] This was not the first milestone of neuroeducation. Various other international events facilitated its emergence as a discipline, such as the first academic course "Mind, Brain, and Education" at Hardvard School of Education, the foundation of the "Australian National Neuroscience Facility", the creation of the "Institut national de la santé et de la recherche médicale" (INSERM), and the Oxford Neuroscience Education Forum in the UK (Pallarés-Domínguez 2015).

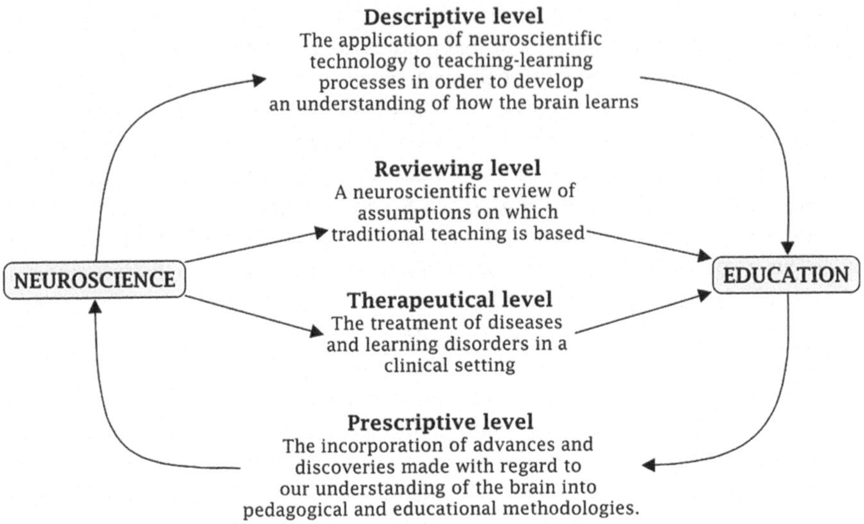

Fig. 10.1 Strata of the relationship between neuroscience and education

of these two areas, both in relation to curricular themes and disorders related to them (Maxwell and Racine 2012; Byrnes 2001).

In spite of these promising advantages, many authors have pointed out the considerable limitations of neuroeducation, three of which are relevant here: (a) asymmetric application; (b) conceptual frameworks; (c) the structural levels and their application. Firstly, neuroeducational experiments have been carried out asymmetrically. In practice, experiments are carried out in what is commonly called "controlled environments", that is, as laboratory experiments, under defined conditions that observe a single person through the use of highly complex equipment. An obvious question arises: is it coherent that neuroeducation studies a single subject when current educational innovation theories emphasise the importance of education as a dialogical process involving several people? Secondly, neuroscience and education use a similar terminology—learning, memory, environment, attention, decision-making, and motivation—but with a very different meaning. Some authors have referred to this problem as one of "category errors" (Davis 2004: 23–27), or the "horizontal problem" (Willingham 2009: 545). Thirdly, a learning-apprenticeship process implies aspects that neuroscience is far from being able to explain, whereby it cannot directly provide an educational policy. The educational process incorporates both external and internal processes. External processes refer to the brain's relationship with others and culture itself. In short, cognition and learning cannot be reduced to neural processes because that would leave out all the other relationships a subject has with their culture, which are not contemplated holistically in neuroscientific studies (Schumacher 2007: 43).

Neuroeducation has made an unprecedented breakthrough, but this is mostly focused on mathematics and language acquisition processes as well disorders related to these spheres of learning. As a result, a series of other questions arise, could neuroeducation make advances in other educational domains? Is moral neuroeducation feasible? Would moral neuroeducation be a new field, or just a form of commentary on the moral dimension of education in general? The term moral neuroeducation is undoubtedly controversial and it needs clarification and justification from a philosophical perspective.

10.3 Moral Neuroeducation: Approaching the Concept

Can neuroeducation, which has been primarily focused on language and mathematics acquisition processes, be extended to other areas? Taking into account the neuroethical discoveries made in the last few years regarding moral judgements, free will, the role of emotions in decision-making, and even the neurobiological basis of altruism or empathy, could we talk about moral neuroeducation (MNE)? If so, should we then apply its results to teaching? Given that the terms that make up the concept of MNE are so wide—moral, neural, education—, it is better to start by defining the relevant concepts.

What is the meaning of moral?[2] According to Spanish philosophical tradition, "ethics" refers to a science that offers reflection on different moral concepts and the diverse ways of justifying moral life in a rational way, thereby gradually guiding human actions. On the other hand, "moral" is a non-necessarily reflective knowledge, whose main components—feelings, values and norms—have been configured socially and historically. Human beings adopt morals freely and consciously by personal conviction, and not in a mechanical way or as a result of their being imposed. The contents of the moral dimension—norms, feelings and values—regulate the relationships between people and between people and the institutions to which they belong. However, for some neuroscientists, the definition provided above is limited, as are other intellectual and philosophical definitions used in "western societies". Instead, neuroscientists have proposed other definitions of morals that are better suited to their experiments. According to Moll et al. (2005):

> Morality is considered as the sets of customs and values that are embraced by a cultural group to guide the social conduct, a view that does not assume the existence of absolute moral values (…). Morality is a product of evolutionary pressures that have shaped social cognitive and motivational mechanisms which had already developed in human ancestors, into uniquely human forms (p. 799).

---

[2] This concept cannot be addressed here in depth. The intention in this chapter is to provide a working definition that enables a conceptualization of moral education. For this purpose, we draw on the Spanish tradition bequeathed by German deontologism, above all by J. Ortega y Gasset, and more recently by J. L. Aranguren, X. Zubiri, and in the synthesis provided by A. Cortina (Aranguren 1997; Cortina 2001; Conill 2003).

On the basis of this definition of morality, it is clear that neuroscience does not take into account the distinctive perspectives drawn between the "ethical" and the "moral" within philosophy; it makes no reference to duty, justice, virtue, or even action. Rather, it is the words emotion, customs, evolution, and social that recur. Needless to say, Moll's neuroscientific definition of morality is not addressed in this study, and it is the philosophical sense of morality that provides a framework for the discussion that follows.

With regard to the concept of learning and the difference between it and education, I draw on the definition and discussion set out by Jesús Conill in chapter one of this volume, and in light of the aforementioned definitions of neuroeducation in the first section, MNE may be identified as referring to the interdisciplinary study of human moral reasoning processes through the consideration of not only ethics, but also neuroscience as a whole. However, it entails more than just studying moral reasoning processes. It seeks to make them explicit in order to enable the creation of moral learning methodologies based on both a knowledge of our brains and ethics. This somewhat general and simple definition raises a series of questions and problematic issues that remain unresolved: (a) the goal of MNE; (b) the concept of moral disorders; (c) the review of previous moral education theories.

First, what is the goal of MNE? In the case of neuroeducation, neuroscientific knowledge is intended to help design practices that enable teaching to be more efficient and to develop a greater intellectual potential in students throughout the course of their education. Therefore, should this also be the goal of MNE? Should we look for moral efficiency? And if so, what is moral efficiency? Are practices that promote survival or well-being morally efficient, as some authors closely tied to neuroethics indeed argue (Churchland 2011)?

Secondly, if neuroeducation deals with educational disorders, should MNE treat moral disorders? However, there are no moral disorders in the same sense as educational disorders.[3] As discussed above, neuroscientific disorders have been applied above all to trying to solve the cognitive learning dysfunctions of dyslexia and dyscalculia. From a moral perspective we cannot talk about moral disorders without falling into the discourse of moral psychology, therefore, we should look for a "bad moral education" instead of "moral disorders" in a psychological context. It is true that, from a psychological perspective, there are cognitive and emotional dysfunctions that have a significant impact on moral behaviour, for example the case of a kleptomaniac, a schizophrenic or a violent psychopath. Moreover, neuroscience has discovered that most of these problems arise from an impairment in some brain areas, such as the dorsal and ventral regions of the prefrontal cortex (PFC), the amygdala, hippocampus, angular gyrus, anterior cingulate and temporal cortex (Oliveira-Souza et al. 2008; Raine and Yang 2006; Blair 2005). It may therefore be concluded that MNE should study other problems, such as political corruption, aporophobia, racial intolerance or gender violence. In other words, it should not try to

[3] Reference is made here to educational disorders understood in the psychological sense, such as dyscalculia or dyslexia amongst others.

alleviate cognitive or emotional dysfunctions that physiologically prevent good ethical behaviour, but rather provide mechanisms to enhance the education of ethical character by taking into account what we know about the brain. After all, is that not what moral education has always attempted? Thereby, the question arises is it necessary to add the prefix *-neuro*?

Thirdly, how can we revise previous approaches to moral education without falling into a 'psy' discourse or neural reductionism? As has been said above, neuroeducation seeks to study or review previous educational theories and verifying their value with regard to what is known about the brain. It is in regard to this concern that the term «neuromythologies» has been coined (Geake 2008), and in accordance with this issue, the third problem faced by MNE is how to verify the most relevant studies and theories on moral education undertaken in recent years. From the most rationalist perspectives of moral education, moral judgement has been considered as a central part of moral action (Brooks et al. 2013). A leading exponent of this relationship between moral reasoning and moral action is L. Kohlberg (1976). Taking into consideration the criticism of Kohlberg's theory of moral stages, especially from C. Gilligan (1982) among others, one of the big challenges faced by his theory concerns the issue of moral motivation (Cortina 2007). In short, if people know what is right, why do they not act accordingly?

Could education provide a way of motivating moral action? If so, how would neuroeducation be able to achieve this without lapsing into a form of neurodeterminism, or else into an attempt at the psychologization of education? Some have seen neuroeducation as granting psychology a still greater role in the pedagogy employed in schools and thus leading to a commodification of social and subjective relationships (De Vos 2016). However, the term MNE is not currently used in academia, and there are a number of theories that could contribute to the development of this concept, and these are discussed below, and a specific focus is developed on the theory of proactive epigenesis.

10.4 Perspectives on Moral Neuroeducation

Teaching is a dynamic phenomenon in which interpersonal interactions occur implicitly and explicitly on many levels. From a neuroscientific point of view, teaching is a cognitive and natural ability of human beings (Lipina et al. 2010). The study of cognitive neuroscience in relation to education has progressively contributed to improving a scientific understanding of education, one that has above all focused on the learner. In contrast, the perspective of other agents and the neural substrates that enable teaching have been overlooked until recently.

The question about whether morality can be taught is one of the oldest educational questions. In fact, one of the initial problems raised in Plato's dialogue "Meno" concerns virtue. Is virtue something natural with which we are born? Alternatively, can it be taught? Current neuroeducation studies have taken up this question and sought to provide new insights through the use of neuroscientific techniques like *functional near-infrared spectroscopy* (fNIRS) (Battro et al. 2013; Holper et al. 2013). According to these studies, the old alternative between "teaching or practice" has been replaced with "learning through action", and the "natural" option now referred to is the "epigenetic processes".

10.4.1 Proactive Epigenesis in Relation to Poverty and Aporophobia

Proactive epigenesis is a current neurobiological approach that explores the existence of an inevitable cultural symbiogenesis in human character formation and behaviour. It examines how human beings can influence the genetic variations of later generations through their character and behaviour. According to this theory, human beings are influenced by their genes and their character in the same way that their actions and their behaviour can, over the long-term, influence the genetic variations of later generations through the stabilisation of synapsis. Awareness and emotion as characteristic human features, as well as the need for coercion, are all addressed by this theory, which aims to account for the genetic transmission of phenotypic traits, that is, the genetic transmission of traits inherited from a being's engagement with its environment. In the case of human beings, the environment includes culture (Evers 2007, 2009, 2015).

Notwithstanding the foregoing, proactive epigenesis could contribute enormously to the potential development of MNE, because it provides a chance for change, not only in our behaviour, but also in the epigenetic trace bequeathed to future generations. Although, two entwined questions remain to be answered—when and how is the epigenetical imprint made in the brain (Salles 2017)—, proactive epigenesis theory has been used to study both social and moral phenomena, such as poverty.[4] Proposals such as these have been articulated through a series of cognitive and developmental neuroscientific programmes, which study how child poverty conditions cognitive development. These programmes have been undertaken in Argentina, in particular over the last decade (Lipina and Segretin 2015: 113; Lipina 2014).

Neuroscientific studies on child poverty conclude that both material and symbolic shortcomings from birth onwards not only limit opportunities for a child's development, but also hinder social inclusion. The impact of any lack in living conditions considered worthy and fair leaves an epigenetic neural imprint that is not homogeneous, nor does it affect people at the same moment in which the state of poverty occurs, instead it manifests itself throughout life.

[4]The question of when and how the epigenetic imprint takes place opens up a debate regarding whether projects that take this theory as a point of departure for studying child poverty, and also use neuroscientific evidence from the same generation of people, are actually studying a pathology or a genetic fingerprint.

Furthermore, poverty affects the configuration and functional regulation of the nervous system at different levels—molecular, hormonal, neural activation, self-regulation and behaviour—and has programmatic effects that imply the neural embodiment of biological, psychological, social and environmental factors (Lipina and Segretin 2015: 113). There have also been several proposals to mitigate the impact of poverty on children. All of these, most of which were developed in Buenos Aires, focus using training to enhance those processes that are central to the cognitive and socio-emotional development of infants who are deprived of primary school and family learning. These processes include attention, inhibitory control, working memory, cognitive flexibility, and phonological processing (Lipina and Segretin 2015: 122). The intention behind the use of these processes related to cognitive improvement is to redress the effects of child poverty, approximately within the first 6000 days of life. However, in so doing it is not intended to produce a moral improvement? So how can they be related to MNE?

Returning to the question of moral motivation, if we know that child poverty is a clear violation of human rights, besides being deeply immoral, why do we continue to allow it? Is it the case that we are not motivated enough to do anything about it? Some have identified this problem as *akrasia—video meliora proboque deteriora sequor—* (Teruel 2016), or as a "radical evil" of the human being (McMullin 2013). For others, the fact that we do not tackle poverty has been understood as "aporophobia" (Cortina 2017). Aporophobia, as a variation of xenophobia, is the rejection of the poor and those who have nothing to contribute in the current society of exchange in which we live. Undoubtedly—and like xenophobia—it is an endemic disease whose neurobiological roots, from an evolutionary point of view, can be traced back to the human being as a dissociative animal.

The emotions that lead to racial and cultural prejudice appear to be based on social emotions, ones that, from an evolutionary point of view, assisted our ancestors in detecting the differences between oneself and "others" that could pose a danger to survival and the welfare of the group. Although these emotions may have obtained questionably "favourable" results for our ancestors, the truth is that it has already been demonstrated, and not only in ethics, but also in economics for example, that these emotions are no longer beneficial. Furthermore, *homo oeconomicus* is much less efficient than *homo reciprocans*, and the extreme version of this, which would lead to selfish individualism, is the result of ideology rather than biological reasoning (Cortina 2017: 72–73; Calvo 2018: 11–14).

A further question remains: does aporophobia have a neurobiological basis? And if it does, could it be modified? Over the last three decades, both neurobiology and evolutionary biology have tried to explain what underpins prejudice and the rejection of both strangers and outsiders. The behavioural tendencies of a group together with the biological foundations of rejection contribute not only to the formation of prejudices, but also to xenophobia (Flohr 1987). More recently, xenophobia has also been studied as a psychopathic and antisocial behaviour due to neural lesions in certain specific areas of the brain, such as frontotemporal dementia. It seems that the mismanagement of moral emotions and the lack of a sense of union between such patients and other people, could explain their antisocial behaviour (Marazziti et al. 2013: 7).

In neuroethics, and more specifically from a proactive epigenesis perspective, the human being has been characterised as an empathic xenophobe by nature. That is, human beings develop an interest in, or reject others on the basis of a selective sympathy (Evers 2009: 121). However, in terms of proactive epigenesis, the fact that humans are empathic xenophobes does not mean they always will be. Our brain is malleable, projective, narrative, creative and autonomously active, it is able to choose between different courses of action and create a genetic imprint that will be transmitted over time through a symbiotic relationship with culture and the environment (Evers et al. 2017). We therefore have the possibility to change, and in turn modify this trend for future generations (Evers 2015: 6).

However, despite being a very interesting debate, this has not addressed the issue of the neurobiological foundation of aporophobia. While, the neurobiological foundations of xenophobic behaviour have been studied, what remains to be examined is the tendency to reject others based simply on the degree of poverty they display? From a moral neuroeducation perspective, the clinical insights on which brain lesions lead to xenophobic behaviour are less relevant than a knowledge about the healthy individuals who developed aporophobic behaviour, and whose behaviour can be reversed through the practice of morally good actions. It is in this context that proactive epigenesis offers a possibility for change. Although the neurobiological foundations of aporophobia remain to be revealed, proactive epigenesis offers us the possibility of reversing this behavioural tendency through interaction with the environment, education and culture.

Therefore, the key factors to consider when answering this question are, firstly, the selection of values and their cultural transmission, and, secondly, how these can result in an epigenetic imprint in the brains of future generations. The epigenesis of neural networks through the stabilisation of a synapse gives rise to the selection of cultural circuits and the transmission of epigenetic cultural traces (Evers 2015: 6–9). And this, in turn, means that cultural traces are a physical reality in the human brain; they leave a physical imprint that is very important in the postnatal period of development and education. Therefore, if we can morally assess, evaluate and choose in our life, we have the responsibility to transmit those values in the form of an epigenetic imprint that can make the world a better place.

In other words, what is so relevant about proactive epigenesis and the stabilisation of brain synapses, is that it creates the possibility for moral education to be based on physiological factors as well as cultural ones. Although we are "empathic xenophobes"by nature—according to epigenetic proactive theory—, our brain offers the possibility for change. Therefore, as humans we are responsible for what we transmit as a genetic inheritance, and this is, a naturalistic responsibility[5] (Evers

[5] Further discussions have taken place in regard to this. Even though proactive epigenesis offers the possibility for change, it does not indicate how to proceed. So how should the supposed "improvement" provided by proactive epigenesis proceed? (Salles 2017). The same problem raises the transformation of the naturalistic fallacy. Proactive epigenesis does not in fact transform the natu-

2009: 148). In this sense, proactive epigenesis provides one of the fundamental neurobiological platforms for MNE studies, as well as a for real educational innovation. The theory examined here suggests that we may be able to influence future evolution by implementing values and norms for a better society (Evers and Changeux 2016: 1363). This implementation of values cannot be done individually, but institutions and social structures would need to assist with the moral development of society (Salles 2017).

10.4.2 Proactive Epigenesis in Relation to Emotional Deprivation

Poverty does not always refer to the lack of material living conditions, but also to other aspects. Another scenario for the application of moral neuroeducation in relation to proactive epigenesis is, for example, the lack of emotional education that turns into emotional deprivation, especially in children. Emotional deprivation is damaging to children primarily due to the absence of an optimal care environment, and this is beginning to be characterised as a new kind of child poverty and a devastating form of child abuse (Brody 1983; Spinley 2007). The UK, for example, has the highest number of middle and upper-class households in Europe where both parents are in full-time work (Pearson 2011). Emotional deprivation can take many different forms, especially with regard to nurturance—when children feel there is no one to support them—, empathy—when there is nobody to listen or understand them—and protection (Young et al. 2011; Kiernan and Huerta 2008).

Emotions are considered to be very complex phenomena formed by multiple factors. They have a powerful influence on people's behaviour and enable them to adapt to their environment. From a neurobiological perspective, emotions designate a group of responses that are activated from the brain to the body and vice versa, and result in the bioregulation of organisms, which enables engagement with complex environmental factors such as society and culture (Damasio 2003). The importance of emotions in learning contexts and how they intensify the activity of neural networks and synaptic connections has been widely studied in the fields of psychology and cognitive neuroscience (Barab and Plucker 2002; Meyer and Turner 2002; Schutz and Lanehart 2002; Mayer et al. 2001). Despite the great effort put into the

ralistic fallacy into naturalistic responsibility but inverts it. However, the term "naturalist responsibility" is problematic. While it states that we are obliged to influence our genetic footprint because we can, it does not tell us how to do this. In other words, it does not philosophically justify the behaviour that should be addressed, or how we must guide this moral action in order to form a genetic imprint for future generations. Therefore, human beings have a responsibility over what they transmit in this form of an inheritance. However, the naturalist responsibility referred to by Evers would only have a moral sense—a true sense of the word responsibility—were it to be complemented by an ethical theory. In earlier studies I have sought to complement proactive epigenesis with the philosophical foundations provided by the Ethics of Cordial Reason (Cortina 2007) for acting morally (Pallarés-Domínguez 2019).

neurobiological study of basic emotions, moral emotions have not been studied to the same extent. From a philosophical point of view, the idea that the human being has a repertoire of emotions with a strong moral component is not new (García-Marzá 2005). Adam Smith, Spinoza, Zubiri and Strawson, all commented on the moral component of emotions. However, moral emotions have been overlooked by the empirical sciences until relatively recently. In fact, the few studies that have been undertaken on the moral emotions have failed to agree on their taxonomy. However, there seems to be some agreement in dividing them into two groups: those that refer to oneself—self-conscious emotions, such as guilt, shame or pride—, and those that we perceive about others—other-conscious emotions, such as gratitude, awe, compassion, disgust, anger or outrage— (Moll et al. 2008; Baez et al. 2017). There has also been a degree of scholarly agreement on the view that moral emotions are the fundamental component of moral sensitivity, which along with the ability to reason and the capacity of judgement form the three fundamental dimensions of human moral cognition (Ostrosky-Solís and Vélez 2008; Marazziti et al. 2013; Narvaez and Bock 2014; Lapsley et al. 2013).

Unlike basic emotions, such as joy or fear, moral emotions relate to an interest or social welfare, and if not to society as a whole, then at least to other people beyond the agent (Haidt 2003). They also contribute to social dissolution or reorganisation, in the form of contempt, outrage and even xenophobia, for example (Ostrosky-Solís and Vélez 2008: 117). Moreover, moral emotions require a more sophisticated cognitive mechanism than basic emotions. They are most likely the result of a combination of subjective emotions, which we possess due to our condition as mammals, and human emotions as well as cognitive mechanisms, in addition to the social emotions we develop as people (Moll et al. 2008: 4).

Without a doubt, the relationship of social emotions with basic emotions is highly remarkable. The neurobiological study of the emotions has addressed this relationship mainly with regard to subjects with brain damage, especially in the ventromedial prefrontal cortex (vMPC), the frontopolar cortex (FPC), the orbitofrontal cortex (OBF), the temporal superior sulcus (STS), in the form of experiments with functional magnetic resonance (fMRI) (Moll et al. 2008: 4; Baez et al. 2017: 183–186). For example, patients with damage to the prefrontal cortex display unusual behaviour in conditions that generate moral emotions such as pride, shame, guilt or repentance (Ostrosky-Solís and Vélez 2008: 120).

Other studies carried out with criminals have revealed the processes by which sociopaths conform with their social environment. In the latter case these processes arise due to alterations in the management of moral emotions, such as a lack of moral sensitivity with regard to empathy, guilt, or repentance, which is caused mostly by external environmental factors, but also by an individual's behaviour (Marazziti et al. 2013: 5), such as drug abuse or a very unstable emotional upbringing at an early age (Narvaez et al. 2016; Brooks et al. 2013). Poverty, in this case emotional poverty, significantly affects the development of people with sociopathic tendencies due to the mismanagement of moral emotions.

Due to the lack of sufficient studies on proactive epigenesis regarding this specific issue, there is no empirical evidence for whether emotional deprivation leaves an imprint on the brain. However, the evidence from previous studies suggests that, like material and socioeconomic poverty, emotional deprivation could have manifest consequences throughout a person's life. Other longitudinal studies have addressed the effects of emotional poverty on the physical brain, for example the case of Romanian children adopted in the United Kingdom, who came from a bleak environment and were completely unaware of emotions (Johnson et al. 1992; Colvert et al. 2008).

Although emotional deprivation has been studied within psychology, a greater focus has been paid to individual treatment and prevention (O'Connor and Rutter 2000), while the idea of a cultural imprint has not been addressed to the same extent. However, studies on children's development have begun to consider epigenetics as a key factor (Child Welfare Information Gateway 2015). According to the National Scientific Council on the Developing Child, environmental and cultural changes "can affect the expression of genes in brain cells, [and] may be permanent or temporary, and can be inherited by the person's offspring" (2010: 3). Although epigenetics has not yet been applied in emotional deprivation studies, it has been addressed in other disorders such as post-traumatic stress disorder (PTSD) (Mehta et al. 2013). Undoubtedly, its application to emotional deprivation would be a promising future line of research.

10.5 Conclusions

As a new discipline, moral neuroeducation needs to undertake a preliminary phase of reflection on its goals, procedures and role. Critical reflection upon MNE must be undertaken from a philosophical perspective in order to address its potential and scope, and also consider how it studies moral disorders and redresses previous moral education theories. One of the most promising neuroscientific approaches to the interaction between brain and education is proactive epigenesis. Indeed, today its potential is not only theoretical, but also practical, because it provides a theoretical framework for the study of the effects that social phenomena, such as poverty and aporophobia, have upon the brain. Over the course of this chapter proactive epigenesis has been studied with regard to its potential use (from a neuroscientific point of view) for moral education, and how it could go on to be studied with in moral neuroeducation in the future. With regard to its application specific emphasis has been placed not only on the poverty of material conditions, but also on emotional poverty, which is understood as the potential detrimental effects of emotional deprivation on children. Proactive epigenesis could be applied to studies on moral emotions and it may potentially shed light on how these emotions condition moral behaviour, the latter being understood as an aspect of human development that evolves over the course of a lifetime.

References

Aranguren, José L. 1997. *Ética*. Madrid: Biblioteca Nueva.

Baez, Sandra, Adolfo M. García & Hernando Santamaría-García. 2017. Moral cognition and moral emotions. In Neuroscience and social science. The missing link, ed. Ibáñez, Agustín, Lucas Sedeño, Adolfo M. García (et al.), 169–197. Cham: Springer.

Barab, Sasha A., and Jonathan A. Plucker. 2002. Smart people or smart contexts? Cognition, ability, and talent development in an age of situated approaches to knowing and learning. *Educational Psychologist* 37 (3): 165–182.

Battro, Antonio M., Cecilia I. Calero, Andrea P. Goldin, Lisa Holper, Laura Pezzatti, Diego E. Shalom, and Mariano Sigman. 2013. The cognitive neuroscience of teacher-student interaction. *Mind, Brain, and Education* 7 (3): 177–181.

Blair, James R. 2005. Applying a cognitive neuroscience perspective to the disorder of psychopathy. *Development and Psychopathology* 17 (3): 865–891.

Brody, Jane E. 1983. Emotional deprivation seen as devastating form of child abuse. *The New York Times*. Available online https://www.nytimes.com/1983/12/20/science/emotional-deprivation-seen-as-devastating-form-of-child-abuse.html. Accessed 19 Mar 2018.

Brooks, Jeff, Darcia Narvaez, and Tonia Bock. 2013. Moral motivation, moral judgement, and antisocial behaviour. *Journal of Research in Character Education* 9 (2): 149–165.

Bruer, John T. 1997. Education and the brain: A bridge too far. *Educational Researcher* 26 (8): 4–16.

———. 2008. Building bridges in neuroeducation. In *The educated brain: Essays in neuroeducation*, ed. Antonio M. Battro, Kurt W. Fischer, and Pierre J. Lena, 43–58. Cambridge: Cambridge University Press. https://doi.org/10.1017/CBO9780511489907.005.

Byrnes, James P. 2001. *Minds, brains, and learning*. New York: The Guildford Press.

Calvo, Patrici. 2018. *The cordial economy. Ethics – recognition and reciprocity*. Cham: Springer.

Campbell, Stephen R. 2011. Educational neuroscience: Motivations, methodology and implications. *Educational Philosophy and Theory* 43 (1): 7–16.

Child Welfare Information Gateway. 2015. *Understanding the effects of maltreatment on brain development*. Washington, DC: U.S. Department of Health and Human Services, Children's Bureau.

Churchland, Patricia S. 2011. *Braintrust. What neuroscience tell us about morality*. Princeton/Oxford: Princeton University Press.

Colvert, Emma, Michael Rutter, Celia Becket, Jenny Castle, et al. 2008. Emotional difficulties in early adolescence following severe early deprivation: Findings from the English and Romanian adoptees study. *Development and Psychopathology* 20 (2): 547–567.

Conill, Jesús. 2003. El carácter hermenéutico y deliberativo de las éticas aplicadas. In *Razón pública y éticas aplicadas: Los caminos de la razón práctica en una sociedad pluralista*, ed. Adela Cortina and Domingo García-Marzá, 121–142. Madrid: Tecnos.

Cortina, Adela. 2001. *Ética aplicada y democracia radical*. Madrid: Tecnos.

———. 2007. *Ética de la razón cordial. Educar en la ciudadanía del s. XXI*. Oviedo: Nobel.

———, ed. 2012. *Guía Comares de Neurofilosofía práctica*. Granada: Comares.

———. 2017. *Aporofobia, el rechazo al pobre. Un desafío para la democracia*. Barcelona: Paidós.

Damasio, Antonio R. 2003. *Looking for Spinoza: Joy, Sorrow, and the feeling brain*. New York: Harcourt Inc.

Davis, Andrew. 2004. The credentials of brain-based learning. *Journal of Philosophy of Education* 38: 21–36.

De Vos, Jan. 2016. *The metamorphoses of the brain. Neurologisation and its discontents*. London: Palgrave Macmillan UK.

Evers, Kathinka. 2007. Toward a philosophy for neuroethics. An informed materialistic view of the brain might help to develop theoretical frameworks for applied neuroethics. *EMBO Reports (Special Issue)* 8: 48–51.

———. 2009. *Neuroéthique. Quand la matière s'éveille*. Luxemburgo: Odile Jacob.

———. 2015. Can we be epigenetically proactive? In *Open MIND*, ed. Thomas Metzinger and Jannifer M. Windt. Frankfurt and Main: MIND Group. https://doi.org/10.15502/9783958570238.

Evers, Kathinka, and Jean-Pierre Changeux. 2016. Proactive epigenesis and ethical innovation. A neuronal hypothesis for the genesis of ethical rules. *EMBO Reports* 17 (10): 1361–1364.

Evers, Kathinka, Arleen Salles, and Michele Farisco. 2017. Theoretical framing of Neuroethics: The need for a conceptual approach. In *Debates about neuroethics. Perspectives on its development, focus, and future*, ed. Eric Racine and John Aspler, 89–107. Cham: Springer.

Flohr, Heine. 1987. Biological bases of prejudice. *International Political Science Review* 8 (2): 183–192.

García-Marzá, Domingo. 2005. ¿Sentimientos virtuosos? El papel de los sentimientos en la vida moral. *Diálogo Filosófico* 62: 241–256.

García-Marzá, Domingo, and Ramón Feenstra, eds. 2013. *Ética y neurociencias: la aportación a la política, la economía y la educación*. Castellón: Servei de Publicacions de la Universitat Jaume I.

Geake, John. 2008. Neuromythologies in education. *Educational Research* 50 (2): 123–133.

Gilligan, Carol. 1982. *In a different voice: Psychological theory and women's development*. Cambridge: Harvard University Press.

Goswami, Usha. 2004. Neuroscience and education. *British Journal of Educational Psychology* 74: 1–14.

Gracia-Calandín, Javier. 2018. Crítica a la naturalización del deontologismo en la teoría del proceso dual del juicio moral de Joshua Greene. *Isegoría. Revista de Filosofía Moral y política* 58: 205–219.

Haidt, Jonathan. 2003. The moral emotions. In *Handbook of affective sciences*, ed. Richard J. Davidson, Klaus R. Scherer, and H. Hill Goldsmith, 852–870. Oxford: Oxford university Press.

Holper, Lisa, Andrea P. Goldin, Diego E. Shalóm, Antonio M. Battro, Martin Wolf, and Mariano Sigman. 2013. The teaching and the learning brain: A cortical hemodynamic marker of teacher-student interactions in the Socratic dialog. *International Journal of Educational Research* 59: 1–10.

Howard-Jones, Paul. 2010. *Introducing neuroeducational research. Neuroscience, education and the brain from contexts to practice*. London: Routledge.

Johnson, Dana E., Laurie C. Miller, and Sandra Iverson. 1992. The health of children adopted from Romania. *JAMA* 268 (24): 3446–3451.

Kiernan, Kathleen E., and M. Carmen Huerta. 2008. Economic deprivation, maternal depression, parenting and children's cognitive and emotional development in early childhood. *The British Journal of Sociology* 59 (4): 783–806.

Kohlberg, L. 1976. Moral stages and moralisation: The cognitive-developmental approach. In *Moral development and behavior: Theory, research and social issues*, ed. T. Lickona, 31–35. New York: Holt, Rinehart and Wilson.

Koizumi, Hideaki. 2008. Developing the brain: A functional-imaging based approach to learning and educational sciences. In *The educated brain: Essays in neuroeducation*, ed. Antonio M. Battro, Kurt W. Fischer, and Pierre J. Lena, 166–180. Cambridge: Cambridge University Press.

Lapsley, D., Anthony C. Holter, and Darcia Narvaez. 2013. Teaching for character: Three alternatives for teacher education. In *The moral work of teaching and teacher education. Preparing & supporting practitioners*, ed. Matthew N. Sanger and Richard D. Osguthorpe, 115–128. New York: Teachers College Press.

Lipina, Sebastián J. 2014. Consideraciones neuroéticas de la pobreza infantil. In *La vida social del cerebro*, coord. Salles, Arleen and Kathinka Evers, 67–101. México: Fontamara.

Lipina, Sebastián J., and Maria Soledad Segretin. 2015. 6000 días más: evidencia neurocientífica acerca del impacto de la pobreza infantil. *Psicología Educativa* 21: 107–116.

Lipina, Sebastián J., Maria S. Segretin, and Maria J. Hermida. 2010. Neuroscience and education: Towards the construction of interactive bridges. *Revista de Neurología* 50 (3): 179–186.

Marazziti, Donatella, Stefano Baronni, Paola Landi, Diana Ceresoli, and Liliana Dell'Osso. 2013. The neurobiology of moral sense: Facts or hypotheses? *Annals of General Psychatry* 12: 6.

Maxwell, Bruce, and Eric Racine. 2012. The ethics of neuroeducation: Research, practice and policy. *Neuroethics* 5: 101–103.

Mayer, John D., Peter Salovey, David R. Caruso, and Gill Sitarenios. 2001. Emotional intelligence as a standard intelligence. *Emotion* 1 (3): 232–242.

McMullin, Irene. 2013. Kant on radical evil and the origin of moral responsibility. *Kantian Review* 18 (1): 49–72.

Mehta, Divya, Torsten Klengel, Karen N. Conneely, Alicia K. Smith, André Altmann, et al. 2013. Childhood maltreatment is associated with distinct genomic and epigenetic profiles in posttraumatic stress disorder. *Proceedings of the National Academy of Sciences of the United States of America* 110 (20): 8302–8307.

Meyer, Debra K., and Julianne C. Turner. 2002. *Educational Psychologist* 37 (2): 207–114.

Moll, Jorge, Roland Zahn, Ricardo de Oliveira-Souza, Frank Krueger, and Jordan Grafman. 2005. The neural basis of human moral cognition. *Nature Reviews Neuroscience* 6: 799–809.

Moll, Jorge, Ricardo de Oliveira-Souza, Roland Zahn, and Jordan Grafman. 2008. The cognitive neuroscience of moral emotions. In *Moral psychology, Vol. 3. The neuroscience of morality: Emotion, brain disorders, and development*, ed. Sinnott-Armstrong Walter, 1–17. Cambridge, MA: MIT Press.

Narvaez, Darcia, and Tonia Bock. 2014. Developing ethical expertise and moral personalities. In *Handbook of moral character education*, ed. Larry Nucci, Darcia Narvaez, and Tobias Krettenauer, 140–158. New York: Routledge.

Narvaez, Darcia, Wang Lijuan, and Ying Cheng. 2016. The evolved developmental niche in childhood: Relation to adult psychopathology and morality. *Applied Developmental Science* 20 (4): 294–309.

National Scientific Council on the Developing Child. 2010. *Early experiences can alter gene expression and affect long-term development: Working paper no. 10.* http://www.developingchild.harvard.edu Accessed 19 Mar 2018.

O'Connor, Thomas G., and Michael Rutter. 2000. Attachment disorder behavior following early severe deprivation: Extension and longitudinal follow-up. *Child & Adolescent Psychiatry* 39 (6): 703–712.

Oliveira-Souza, Ricardo, Robert D. Hare, Ivanei E. Bramati, Grisela J. Garrido, Fatima Azevedo Ignàcio, Fernanda Tovar-Moll, and Jorge Moll. 2008. Psychopathy as a disorder of the moral brain: Fronto-temporo-limbic grey matter reductions demonstrated by voxel-based morphometry. *NeuroImage* 40 (3): 1202–1213.

Ostrosky-Solís, Feggy, and Alicia E. Vélez. 2008. Neurobiología de la sensibilidad moral. *Revista Neuropsicología, Neuropsiquiatría y Neurociencias* 8 (1): 115–126.

Pallarés-Domínguez, Daniel. 2015. Hacia una conceptualización dialógica de la neuroeducación. *Participación educativa Revista del Consejo Escolar del Estado* 4 (7): 133–141.

———. 2016. Neuroeducación en diálogo: neuromitos en el proceso de enseñanza- aprendizaje y en la educación moral. *Pensamiento* 72 (273): 941–958.

Pallarés-Domínguez, Daniel, and Andrés Richart. 2018. Entre la neuroética y la neuroeducación: las fronteras de las neurociencias sociales. *Recerca. Revista de Pensament i anàlisi* 22: 7–13.

Pallarés-Domínguez, Daniel. 2019. La construcción dialógica de una neuroética filosófica. In *Ética y Democracia desde la razón cordial*, ed. Elsa González, Juan C. Siurana, José L. López-González, and Marina García-Granero, 169–175. Granada: Comares.

Pearson, Allyson. 2011. Britain's middle-class children are so privileged – and so neglected. *The Telegraph*. Available online https://www.telegraph.co.uk/women/mother-tongue/8850638/Britains-middle-class-children-are-so-privileged-and-so-neglected.html. Accessed 19 Mar 2018.

Raine, Adrian, and Yaling Yang. 2006. Neural foundations to moral reasoning and antisocial behaviour. *Social Cognitive and Affective Neuroscience* 1 (3): 203–213.

Salles, Arleen. 2017. Proactive epigenesis and ethics. *EMBO Reports* 18 (8): 1271.

Schumacher, Ralph. 2007. The brain is not enough: Potential and limits in integrating neuroscience and pedagogy. *Analyse und Kritik* 29: 38–46.

Schutz, Paul A., and Sonja L. Lanehart. 2002. Introduction: Emotions in education. *Educational Psychologist* 37 (2): 67–68.

Sigman, Mariano, Marcela Peña, Andrea P. Goldin, and Sidarta Ribero. 2014. Neuroscience and education: Prime time to build bridge. *Nature Neuroscience* 7 (4): 497–502.

Spinley, B.M. 2007. *The deprived and the privileged: Personality development in English society*. London: Routledge.

Teruel, Pedro J. 2016. Deteriora sequor. Interpretación neurofilosófica del fenómeno de la akrasía. *Pensamiento* 72 (273): 865–880.

The Royal Society. 2011. *Brain waves module II. Neuroscience: Implications for education and lifelong learning*. London: The Royal Society Science Policy Centre.

Tokuhama-Espinosa, Tracey N. 2011. *Mind, brain and education science. A comprehensive guide to the new brain-based teaching*. New York: Norton & Company.

Willingham, Daniel T. 2009. Three problems in the marriage of neuroscience and education discussion. *Cortex* 45 (5): 544–545.

Young, Robert, Susan Lennie, and Helen Minnis. 2011. Children's perceptions of parental emotional neglect, control, and psychopathology. *Journal of Child Psychology and Psychiatry* 52 (8): 889–897.

Chapter 11
The Case of Gender in Moral Neuroeducation

Sonia Reverter-Bañón

11.1 Introduction

The principal thesis of this chapter is that neuroeducation must engage with a critical concept of gender in order to avoid political decisions being taken that would reinforce the regulation of life according to a simplistic binary view of gender. Thereby, neuroeducation can become an ally in tackling one of the most prominent neuromyths[1] (Geake 2008). To address these concerns three key issues must be addressed:

(a) ascertain what is understood by the concepts of sex and gender, above all with regard to the consequences of neuroeducation;
(b) survey the state of research concerning moral cognition and sex/gender in order to better understand the conceptual framework structuring the practice of moral neuroeducation with regard to sex/gender;
(c) present a series of conclusions on future directions to be taken in research on moral neuroeducation.

[1] Geake (2008, 123) defines neuromyths as "those popular accounts of brain functioning, which often appear within so-called 'brain-based' educational applications". As Geake (2008, 124) states the problem is that "the evidential basis of these schemes does not lie in cognitive neuroscience, but rather with the various enthusiastic promoters; in fact, sometimes the scientific evidence flatly contradicts the brain-based claims".

S. Reverter-Bañón (✉)
Universitat Jaume I, Castellón de la Plana, Spain
e-mail: reverter@uji.es

P. Calvo, J. Gracia-Calandín (eds.), *Moral Neuroeducation for a Democratic and Pluralistic Society*, https://doi.org/10.1007/978-3-030-22562-9_11

11.2 Concepts of Sex and Gender in Neuroscientific Research

11.2.1 The Neuromyth as a Scientific Thesis

Since the term neuroeducation was first defined by Battro and Cardinali (1996), the discipline has grown exponentially. However, despite the seeming novelty of the term applied to this new discipline, there had been previous attempts to find possible applications of neurobiological findings for education. As Theodoridou and Triarhou (2009) note, the work of two men, the neurologist Henry Herbert Donaldson (1857–1938) and the educationist Reuben Post Halleck (1859–1936), can be considered as pioneering in this field. Donaldson published his book *The Growth of the Brain* in 1895, and Halleck's *The Education of the Central Nervous System* came out a year later in 1896.

Both authors examined what continue to be keenly-debated issues in neuroeducation today: the plasticity of the human brain, the "nature versus nurture" debate, and teaching and learning strategies for optimising education. Despite their differences these two authors' ideas can be taken as a point of departure that frames the core interest of neuroeducation: to understand how the brain and education interact in a two-way dialogue in which neither prevails over the other (Ansari et al. 2012; Bruer 1997, 2008).

Both Donaldson and Halleck explored the question of physiological differences in male and female brains and highlighted the need for research on the impact had by the environment, nurture and education on reinforcing brain structures and stimulating cognitive abilities. However, it is Edward H. Clarke's (1820–1877) book, *Sex in Education* (1874), that must be considered as the true pioneer in establishing a connection between sex and education. Its publication caused huge controversy among feminist activists, who at that moment were gaining collective strength through the movement for women's equality and its role in demanding for a right to vote and an equal civic status.

Clarke's main line of argument is that women's health is incompatible with intellectual activity:

> Woman seems to be looking up to man and his development, as the goal and ideal of womanhood. The new gospel of female development glorifies what she possesses in common with him, and tramples under her feet as a source of weakness and badge of inferiority the mechanism and functions peculiar to herself. (Clarke 1874: 129)

Similar arguments were routinely employed throughout the nineteenth and twentieth centuries. Their core tenet being that co-education causes physiological harm to women, and this view was repeated by various scientists over the course of the twentieth century. In 1915, for instance, the renowned neurologist Charles Dana expressed his concern that were women to live equally alongside men, they would run the risk of insanity (in Fine 2010: 131). His opinion was founded on a physiological basis, namely that the upper half of the spinal cord, which controls the pelvis and the limbs, is smaller in women. According to Dana this affects women's efficiency in evaluating political initiatives or exercising judicial authority.

A number of feminist activists were quick to respond to Clarke's physiological arguments, and the same year it was published, 1874, *Sex and education: A reply to Dr. E.H. Clarke's "Sex in education"* was also issued. Edited and with an introduction by Julia Ward Howe, this book set out to challenge Clarke's thesis. The first argument that Howe proposed may today be read as pre-empting the contemporary concern for the need for dialogue between education and the biological and human sciences:

> To those most eminent in physics and in sociology we would say: "Take the social mixture of to-day, with its antecedents and concomitants. Analyze it fairly and thoroughly; and then tell us if the over-education of women is its most poisonous ingredient". (Howe 1874: 7)

Howe undertook a detailed critique of each of Clarke's arguments on the danger and inappropriateness of co-education as well as the teaching of gender equality. Her substantial conclusive argument is as follows:

> It therefore seems to me that, as his facts are not worked out with sufficient thoroughness to justify any general conclusion whatever, so his conclusion that our present colleges, and particularly Harvard College, cannot, except at a vast expense, admit women, is utterly unsustained by his facts. (Howe 1874: 21)

Our question today, more than a century later, is: are there scientific arguments that can justify the thesis of gender differentiated education? However, such a line of enquiry clearly calls for an understanding of what the aim of education is. Again, recourse can be made to Howe, whose fundamental ideal of education continues to be valid today:

> [...] education is not merely the knowledge of sciences, languages, or systems of philosophy, but consists in the use of the faculties and their application to life thus developed by these and other studies. (Howe 1874: 24)

Roskies, in her renowned article "Neuroethics for the New Millennium", argued that an understanding of the brain mechanisms underpinning human behaviour has "potentially dramatic implications for our perspective on ethics and for social justice" (2002: 21). Therefore, neuroeducation must not only undertake research on and debate issues regarding the knowledge provided by different academic disciplines, but also address the entwined moral questions. It is in the latter context that the question of the differences and similarities between male and female brains is essential, yet prior to any discussion of neuroeducation or moral neuroeducation a clear understanding of the concepts of sex and gender is a fundamental prerequisite. The historical examples outlined above reveal the confusion resulting from discussion of the brain's physiology. Examples such as these conflate physiological and behavioural arguments, and they highlight the need to refine the concepts of sex and gender that underpin them (Hyde 2007, 2016; Lorber 1996; Johnson and Repta 2012; Runnels et al. 2014).

11.2.2 Concepts of Sex and Gender

As a point of departure, the idea of gender addressed here is as a concept that refers to a complex phenomenon deriving from mechanisms of inequality (Repo 2015). Thereby, it is argued that gender be understood as a concept of the parallel construction of male and female identities within the system of patriarchy. Patriarchy organises subjectivities in relation to the diverse modes of power that establish meaning in human communities. It is in regard to this patriarchal structure that diverse feminist theorists have identified gender as a construction of inequality. The implementation of gender is linked to the concept of sex, whereby the concepts of each feed into one another. The Möbius strip offers an insightful visual metaphor of the relationship between the two concepts in the context of identity construction (Grosz 1994). Thereby, the arguments that follow take up the proposal made by various feminist theorists that the concepts sex/gender should be addressed together as a single phenomenon.

By doing so this approach seeks to clarify the ongoing confusion between the concepts of sex and gender with regard to brain difference. Following the development and circulation of the concept of gender during the period of second-wave feminism, a conceptual ambiguity has persisted that hinders a coherent understanding of the discourse on sexual difference. Maccoby and Jacklin's classic study, *The Psychology of Sex Differences* (1974), sought to overcome this confusion by taking an interactionist position that aspired to explain the few differences found in cognitive processes. Five years later in *Toward a Redefinition of Sex and Gender*, Unger (1979) called for a redefinition of research on sexual difference to clarify the distinction between sex and gender, and how these terms should be used. Unger (1979: 1085) clearly states, "A major problem in this area appears to be the too inclusive use of the term sex". In other words, the term sex is ambiguous, as it is physiological, functional, cognitive, cultural, social, and so on. And this is what underlies the problem that still gives rise to confusion between the two terms in some examples of scientific research.

In recent years some members of the scientific community[2] have criticised the ongoing confusion between sex and gender in the neuroscientific literature, along with what has been termed "the production of ignorance" (Spelman 2007) and

[2] We refer mainly to the NeuroGenderings Network (NGN), which was the outcome of a conference in 2010 funded by the Swedish Research Council and organised by the Center for Gender Research at the University of Uppsala. The motivation for the conference, entitled *NeuroGenderings: Critical Studies of the Sexed Brain*, was to help distinguish scientific research from pseudo-science in the area of male-female brain differences. The 2010 event was followed by two further conferences (Vienna 2012, and Lausanne 2014). The NGN is an international transdisciplinary group of scholars whose aim is "[…] to critically examine neuroscientific knowledge production and to develop differentiated approaches for a more gender adequate neuroscientific research. Feminist neuroscientists generally seek to elaborate the relation between gender and the brain beyond biological determinism but still engaging with the materiality of the brain" (www.neurogenderings.wordpress.com).

"epistemologies of ignorance" (Tuana 2004, 2006). The combination of these confusing issues provides a clear example of the construction of non-knowledge, which in turn leads to the creation of the "neuromyth" associated with sexual difference (Reverter-Bañón 2016, 2017).

The theoretical proposal set out here, and formulated as a critical perspective, is not intended to untangle this discursive mêlée merely by separating sex and gender, but instead it seeks to establish an explanatory framework within which they can be combined and related to one another without confusion. The renowned biologist Fausto-Sterling (2000) has proposed using the theory of dynamic systems to develop our understanding of the sex/gender relationship, as this theory provides a more appropriate and productive framework than the traditional nature-nurture dichotomy. In line with other feminist thinkers (Butler 1990; Spivak 1993; Grosz 1994), Fausto-Sterling (2000) argues that sex and gender are both partially social constructs, but because they are materialised in the body they are simultaneously biological. The theory of dynamic systems allows us to combine the social and the biological in order to understand how cultural difference becomes biological difference. Fox Keller (2010) likewise has argued for the need to go beyond the nature-nurture dichotomy in order to find a language that better reflects contemporary science and helps us find a way "out of the morass in which we have been bogged down for so long" (Fox Keller 2010, 13).

The blurring between sex and gender in the current neuroscientific literature has reached such a state that the differences between men and women are referred to as "sex differences" in some cases and "gender differences" in others, irrespective of whether they are neuronal or behavioural. For example, Kaiser (2012: 131) has highlighted the persistent terminological confusion in leading neuroscientific publications such as *Cerebral Cortex* and *Nature Neuroscience*. One example cited by Kaiser is how one multi-authored paper (Gur et al. 2002), published in the former journal, refers to *sex differences* in the frontal brain volumes, while in a similar paper (Lüders et al. 2004), published in the latter journal, *gender differences* are identified in the brain's cortical complexity. As Kaiser notes, this indecision about the correct terminology reflects a deeper uncertainty about how to observe and explain the origin of differences between men and women. Schmitz (2012) has also observed how this indecision implies a degree of uncertainty and/or disagreement as to whether male and female behavioural differences are attributable to the brain's sex, or whether these brain differences can be explained by differences in behaviour.

By proposing to interrelate sex and gender in this study it seeks to foreground how the study of the body is by no means restricted to the sphere of the biological sciences. Bodies are formed, transformed and function within a discursive framework that gives rise to interpretations of what they are. Cultural and social meanings are inscribed within the human body. As gender is one of the most apparent identity inscriptions, it is of key importance to review how biology and culture operate together to implement gender identity, and it is especially pertinent in order to steer or guide the discourse within neuroeducation on sex/gender differences. We need to consider the ways in which a neuroscientific framework can have significant theoretical and ethical consequences for children's education. It is therefore crucial to observe whether this framework gives rise to essentialist concepts of men and women. The illusion that male-female brain differences are biologically innate has been heavily criticised (Fine 2012; Halpern et al. 2011; Vidal 2012), as it leads to the view that the differences between men and women are inculcated from birth and are, therefore, immutable. However, there is evidence to suggest that an integral understanding of gender requires explanations at multiple levels, ones that encompass the genetic and the physiological, as well as the social and cultural.

A neuroeducational approach that does not critically examine the concept of sex/gender brain differences may give rise to mistaken assumptions in the education of girls and boys. It may perpetuate or even encourage gender-biased curricula and result in decisions being taken with regard to education that hinder the provision of equal opportunities for educational success (Sadker and Zittleman 2009). Rather than tackling gender-based inequalities in the education system, therefore, the provision of education based on neuroscience could foreseeably shape decisions that result in the legitimisation of such inequalities. Neuroeducation must therefore give much greater consideration to this ethical debate (Busso and Pollack 2015).

11.3 Moral Neuroeducation in Relation to Sex/Gender

If moral education is addressed within the context of the findings of the neurosciences—what this research group terms "moral neuroeducation"—any decisions taken in the absence of an adequate critical framework, in the sense addressed above, may prove to be an obstacle towards achieving an egalitarian society. Urgent reflection is therefore needed on the issue of gender within moral neuroeducation.

To this end, an overview of the state of research on this issue is addressed, following which a series of approaches to achieving equality are proposed. The core argument of this chapter is that the concept of "sex/gender" denotes a multidimensional reality and is therefore a dynamic and relational phenomenon. The study of sex/gender identity must therefore engage with: (a) the material conditions underpinning whether or not we can speak of a natural or biological difference between male and female brains; (b) the economic, social and political factors that arise in connection with inequality on the grounds of sex/gender identity.

In addition, this multidimensional approach is relational, which means that the aforementioned material conditions and diverse factors are interrelated, and as feminist theory has forcefully argued in recent decades through the concept of intersectionality (Crenshaw 1991), this is a key issue for understanding the mechanisms of exclusion and inequality. By undertaking this enquiry on the significance of sex/gender identity it seeks to underscore its relevance for neuroeducation research, and in regard to this, attention will be paid to issues of the body, biology and the brain, as well as social, economic and political concerns that must be considered as an integral part of the dialogue. On the basis of this discussion what this chapter proposes is that the study of the sex/gender variable in education is a key line of research that permits the study of the development of two concepts that have advanced notably in neuroscience over the last decade: brain plasticity and epigenesis (Evers 2015). By taking these two concepts into account a deeper understanding can be gained of how nature and nurture interact, which is fundamental for understanding the sex/gender combination. Furthermore, this approach enables us to explore specific possibilities neuroeducation's contribution to the establishment of equality within education.

11.3.1 Proposals for Integrating the Gender Perspective into Moral Neuroeducation

To a certain extent, the discussion undertaken on the pioneering ideas about sex/gender brain differences in the first part of this chapter, what we now call neuroeducation, reveals that the crux of the discussion revolves around the nature-nurture debate. A survey of developments in this debate reveals that it is above all related to the question of sex brain differences. In other words, the nature-nurture debate can be traced in parallel to the debate on sexual difference,[3] and in recent decades with the rise of the neurosciences, to the debate on male-female brain differences.

In neuroeducation this debate has largely centred on the question of whether boys and girls should be educated together or separately (Eliot 2011, 2013; Liben 2015). For decades, the research on intervention programmes that might reverse patterns of gender inequality has focused on improving opportunities for the development of girls and young women. The design of these programmes calls for a multiple intervention that addresses the complexity of the construction of sex/gender identity. The challenge of overcoming inequality has particularly centred on the cultural and social spheres, in which education plays a considerable part. Egalitarian patterns for gender construction aim to prevent the reproduction of identity roles that highlight and foster differentiated capabilities for men and women, thereby empowering the former and disempowering the latter. Neuroeducation, and more specifically moral neuroeducation, can play a key role in this endeavour. The challenge today in ascertaining whether there are sex differences in the brain, and at the same time whether they are innate or a product of our environments. Over the course of the next section we review a range of research that helps clarify this question.

[3]As Cynthia Russett (1989) pointed out in her classic book *Sexual Science: The Victorian Construction of Womanhood*. Russet argues that the nature-nurture debate is largely structured through the study of sex differences.

11.3.1.1 Studies on Sex Differences in Moral Cognition

Neuroscientific research on sex differences in moral cognition is still scarce. The few studies that have explored this question are far from conclusive and highlight the need to undertake brain research at a deeper level. The pioneering study to date in this field, although not strictly neuroscientific research, is Jaffee and Hyde's "Gender Differences in Moral Orientation: A Meta-Analysis" (2000), which provides a quantitative review of empirical studies devoted to the issue of gender difference and moral orientation.

Jaffee and Hyde's meta-analysis is based on Carol Gilligan's (1982) theory of cognitive moral development, in which she advocated an extension of what we understand by moral reasoning. Gilligan's work responded to Kohlberg's studies (1976, 1984) on states of moral development, which she claimed were based on an exclusively justice-oriented idea of moral reasoning, and ignored the role of care. Kohlberg argued that the level 3 of moral reasoning (characterised by an interest in human relationships and meeting others' expectations) may be identified with the female gender, while level 4 (characterised by a law and order mentality concerned with ensuring social order) can be identified with the male gender. In her critique of Kohlberg, Gilligan argued firstly, that care-oriented moral reasoning is not inferior, and secondly, that if one type of moral reasoning (justice as opposed to care) is found more consistently in one sex than another, this is due to nothing other than the dichotomised gender differences integrated into children's upbringing. After decades of debate on the respective positions of Kohlberg and Gilligan (Walker 2006), Jaffee and Hyde's (2000: 707) meta-analysis reveals practically no sex/gender differences in moral reasoning. They (2000: 719) conclude that: "[…] although distinct moral orientations may exist, these orientations are not strongly associated with gender".

Jaffee and Hyde's meta-analysis, which covers the period 1966–1998, focuses on two types of study, those employing interview measures and those based on objective tests. However, it was during the 1990s, and with an ever-increasing recurrence in the twenty-first century, that new brain study techniques have allowed researchers to design experiments to "observe" sex differences in the brain, and also equipped them with neuroimaging techniques to analyse these possible brain differences. However, the use of these methods and techniques can be tainted with the prejudices identified above. For instance, Bluhm's (2011, 2013) work on the influence of gender stereotypes in research on emotions that uses functional magnetic resonance imaging (fMRI), highlights the need for caution, given that "[…] its conclusions about sex differences owe more to gender stereotypes than to evidence" (Bluhm 2013: 870).

Among the studies on sex differences in moral cognition, the following are the most significant: Robertson et al. (2007), Harenski et al. (2008), Fumagalli et al. (2010a, b) and Scheele et al. (2014). Robertson et al.'s (2007) study uses fMRI to study real life problems with the aim of:

> [isolating] the neural correlates of one specific component of moral thought, the ability to recognize and interpret moral issues (i.e., moral sensitivity), and [comparing] the neural processing related to moral sensitivity to two different types of moral content represented by issues of justice and care. (Robertson et al. 2007: 756)

Yet although this study shows that moral sensitivity is associated with certain brain areas, it does not serve our purposes as it only included 16 male subjects. As the authors acknowledge (Robertson et al. 2007: 763), the study needs to be replicated with female subjects so that possible sex/gender effects may be identified in the neural correlates of care and justice on moral judgements.

Exactly this replica experiment was performed by Harenski et al. (2008). Their study, "Gender differences in neural mechanisms underlying moral sensitivity", used fMRI to examine the brain regions implicated in moral judgement and gender differences in moral sensitivities. Their study found no differences between male and female brain activity while the participants viewed pictures; however, differences were observed in brain activity when the subjects rated the degree of moral violation. The women's brains showed more activity in areas associated with emotion (posterior and anterior cingulate, interior insula), while the men's brains showed more activity in regions implicated in cognitive processing (superior temporal sulcus). The authors concluded that their data was consistent with the belief that women approach moral dilemmas from a care-based orientation, while men tend to adopt justice-based moral evaluations (as Carol Gilligan had posited in 1982):

> In summary, males may have used executive resources to evaluate multiple contextual aspects of pictures in evaluating violation severity, whereas females may have focused more on the perception of individuals in distress, e.g. the target of a moral transgression (Harenski et al. 2008: 320).

Then, in a study of 78 healthy subjects (38 men and 40 women), Fumagalli et al. (2010a) found that direct current stimulation of the ventral prefrontal cortex increased moral judgements that allowed personal harm to be inflicted in order to increase aggregate welfare. Subsequently, another study was undertaken by Fumagalli et al. (2010b), which was based on Greene's (2009) dual process theory, according to which personal moral actions (that activate the affective system) are separate from impersonal moral actions (that activate the cognitive system). Fumagalli and her colleagues claimed that their findings, based on moral dilemma tests, suggest that:

> [...] the cognitive–emotional processes involved in evaluating personal moral (PM) dilemmas differ in men and in women, possibly reflecting differences in the underlying neural mechanisms. Gender-related determinants of moral behaviour may partly explain gender differences in real-life involving power management, economic decision-making, leadership and possibly also, aggressive and criminal behaviours. (Fumagalli et al. 2010b: 219)

Fumagalli et al.'s (2010b) analysis suggests that in the case of personal moral (PM) dilemmas, men are more likely than women to make utilitarian judgements (in which the morality of an action depends on its consequences PM) than deontological judgements (in which the morality of an action depends on its coherence with moral norms).

Like Fumagalli (2010b), Scheele et al.'s (2014) experiments with healthy adults (stage 1: 23 men; stage 2: 74 men and 60 women) were based on Greene's dual process theory.

> These authors concluded that oxytocin (OXT), applied intranasally, increased moral endorsement in men but did not affect women: We argue that increased self-focus evoked by OXT is compatible both with enhanced processing of salient negative or ambiguous emotional cues and also with increased selfishness to promote survival both of self and of an individual's own family group. On the other hand, in women, OXT may act upon enhancing other regarding behavior to facilitate offspring survival by promoting altruistic decisions and wider social group cooperation, perhaps reflecting the importance of communal care. (Scheele et al. 2014: 6074)

All the above-mentioned experiments were conducted with adult subjects and therefore shed no light on whether the differences detected are innate or not. Scheele et al. (2014: 6074) cautioned that "it is currently unclear whether sex-specific response patterns remain stable across the lifespan, precluding an extrapolation of our findings to aged populations". Similarly, Fumagalli et al. (2010b: 223) noted that "Whether gender differences in behavioural measures arise from cultural effects or reflect innate differences remains unclear".

Although the recent meta-analyses by Bryant et al. (2016) and Garrigan et al. (2016) on the neural correlates of moral decisions do not include the sex/gender variable in their analysis, they do inform us of the limitations of neuroscientific research on moral decisions, and they highlight the difficulty of distinguishing when behavioural differences are the consequence of neurological differences (Bryant et al. 2016: 38). In other words, the overriding question of whether the few small differences detected are innate or due to environmental influences remains unanswered.

Garrigan et al.'s (2016) meta-analysis cautions that because all the studies on neural correlates were conducted on adult subjects, no conclusions can be drawn about whether differences may change over the life course:

> The conclusions drawn, therefore, only apply to adults. Further neuroimaging studies focusing on children and adolescents would help answer questions about moral development and the developmental pattern of the neural correlates of moral decisions. (Garrigan et al. 2016: 96)

Neurologist Catherine Vidal (2012) warns that when the conclusions of studies such as Harenski et al. (2008), suggesting that women tend to approach moral dilemmas with a care-based orientation and men with a justice-based orientation, are disseminated to the general public, they generate a belief that men are more reliable than women when making moral judgements because they have a more rational brain. This conclusion, however, has no grounds in Harenski et al.'s (2008) work.

According to Vidal (2012: 300), in relation to scientific studies and their conclusions, three points must be made clear:

(a) Brain differences in a small sample of participants are not statistically significant, whereby it is quite plausible that if a larger sample were analysed, sex differences would disappear due to the high inter-individual variability in brain functioning. Robertson et al. (2007) noted this limitation in their own study.

(b) fMRI data are only seen in a fixed image of a given state in the brain of one individual. They provide no direct evidence on the biological factors or socio-cultural processes that have influenced this state. An image of sex differences in the brain's structure or function does not imply that the differences exist at birth, nor that they would be recorded throughout the individual's life. As we have seen, the aforementioned authors recognise the limitations of their studies in this respect.

(c) Their findings were obtained in an artificial laboratory context, which cannot be extrapolated to real experiences of making moral judgements. In their meta-analysis, Garrigan et al. (2016: 96) highlight the need for "real life scenarios" to extrapolate the findings of these experiments: "We recommend that future neuroimaging experiments use more real life scenarios for assessing moral decision-making, for example, everyday scenarios that people are more likely to encounter than life or death situations" (Garrigan et al. 2016: 96).

As Bluhm (2013) concluded in her analysis on the influence of gender stereotypes in emotion research undertaken using neuroimaging techniques, the conclusions reached are a self-fulfilling prophecy. Indeed, there is a trend, that has in recent years been repeatedly singled out for criticism, to draw illegitimate conclusions about sex/gender differences, and this has been illustrated in the study by Fumagalli et al. (2010b), who clearly state that it is impossible to say whether the differences they detected are innate or due to cultural factors. Despite this, they nonetheless affirm that:

> Unlike gender, cultural factors such as education and religious belief had no effect on how the university students we studied performed the moral judgment task. Hence, we conjecture that gender differences are probably better explained by biological than by cultural differences. (Fumagalli et al. 2010b: 223)

As they make clear this is a conjecture, not a conclusion, yet it is granted a degree of authority through being published in a scientific study endorsed by the rigour of academic publishing. Furthermore, due to the powerful influence exerted by neuroscience on the general public, such views are highly likely to appear in the mass media as a conclusion and not a conjecture (Maney 2016).

A recent example of this phenomenon is a press article written by the neurologist Pullicino, entitled "Do genders have same brain?" (Times of Malta, 12-1-2014).[4] In the article, which appeared in the context of proposed sex change legislation in Malta, the author drew on several neuroscientific studies, including that of Fumagalli, to state:

> Research is telling us we have a hard-wired brain structure that underlies the sex-specific way we act. We can change the external way we act and look into that of the opposite sex, but it is likely that a lot of the actions and decisions of transsexual people will be strongly influenced by the brain structure of their original sex.

[4] Available at: https://www.timesofmalta.com/articles/view/20141201/opinion/Do-genders-have-same-brain-.546355 (last accessed 28-10-2018).

11.4 Conclusions and Future Directions

Current neuroscientific knowledge is insufficient to endorse the implementation of education programmes that differentiate according to sex/gender, and as we have seen, nor can it provide evidence for differentiated patterns of moral judgement. The long history of neuromyths, confusions and prejudiced interpretations provide a clear warning against any further deceptions arising from pseudoscientific declarations on sex/gender brain differences (Fine 2010). Given how easy it is to misinterpret scientific theories and experiments on sex brain differences (Fine 2010; Jordan-Young 2010; Bluhm 2013), new guidelines on the design of experiments and their procedures should be introduced to prevent essentialist prejudices.

Along these lines, Rippon et al. (2017) in a letter to the *Journal of Neuroscience Research*, made an interesting proposal by calling for the appropriate consideration of sex as a biological variable. Some of the co-authors of this letter had previously published their ideas on how this should be done in the article "Recommendations for sex/gender neuroimaging research: key principles and implications for research design, analysis, and interpretation" (Rippon et al. 2014). On the basis of their research on the influence of sex differences on the brain they recommend publishing not only the differences found, but also any failures to replicate the expected differences and similarities. As Maney (2016: 282) points out "[…], because sex differences are so easily misunderstood and misinformation potentially harmful, we need to hold others and ourselves to a high standard when reporting them".

Neuroeducation, and more specifically moral neuroeducation, must be highly cautious in transferring neuroscientific knowledge to the classroom and to education in general (Derks and Krabbendam 2013). It is precisely because of both the errors noted and the ambitious promises made in brain research that the new discipline of neuroeducation must be developed to ensure that an interdisciplinary dialogue and two-way exchange between neuroscience and education is a permanent *modus operandi* in this field of study (Busso and Pollack 2015; Bruer 2008). At the same time, both morality and gender are complex phenomena in themselves, and the question of how they interact is challenging to study unless a comprehensive perspective is taken (Walker 2006).

The neurosciences have the potential to further our understanding of social problems such as inequality. Neuroscientific knowledge could enhance the design of educational programmes to prevent exclusion and inequality, but this will not happen without a rigorous interdisciplinary dialogue that is committed to such possibilities for social transformation. Future neuroscientific research will need to extend and develop an enquiry into the impact of upbringing on the human brain. In regard to, this undoubtedly, the effects of unequal male and female gender construction on the brain, may reveal unexpected and innovative aspects, above all due to the profound and far-reaching impact of gender construction on the creation and regulation of human life. Scientific research on these issues may prove to be key when designing education programmes intended to make our societies more egalitarian.

At present, our understanding is that moral neuroeducation can help to make these issues into an area of fruitful research for the future.

In setting research on these issues as a goal for moral neuroeducation, as proposed here, it should be highlighted that there is a need to evaluate individual differences, and not only those that can be detected in sex/gender identity groups. Holtzman's (2017) study on neuro-moral diversity, is especially relevant in this regard, above all in relation to his call for caution regarding the concept of the "moral brain":

> The moral brain typically refers to a theoretic construct that is partly premised on the idea that by studying how some brains formulate moral judgments, we can come to understand how brains in general do. (Holtzman 2017: 1)

His call for caution is especially relevant in the case of research on male and female brains since individual diversity may be clouded by the design of neuroscientific research that divides all individuals into sex/gender groups.

Also relevant to this is what Kraus (2011) terms "neurodiversity"; neurodiversity refers to the idea that there is a natural variation among brains, that this variation is inherently good for humanity, and that societies must understand this, and value it as an asset. Critical perspectives such as this fall within what Kaiser and Dussauge (2012, 2015) term "non-neuro-normative" discourses, which advocate a critical and transdisciplinary approach to what the brain is. In other words, knowing what counts as male-female sex difference is highly relevant. Chalfin et al. (2008: 2) state this very clearly:

> Although the answer to this question may appear self-evident, history shows that our answer is subject to the constraints of time and place. Feminist and social studies of science have demonstrated how scientific research, questions, and constructs reflect contemporary cultural presuppositions and beliefs about gender.

Views such as this raise the need for a thorough review of the conceptual framework within which neuroscientific research takes place, and some members of the scientific community are already undertaking such a review (Bluhm et al. 2012). In a context where partial, and even prejudiced or simply pseudoscientific research has a long and deep-rooted history marked by a persistent recurrence, brain neuroscience needs an ethical commitment that embraces what some have termed "critical neuroscience" (Choudhury and Slaby 2012; Schleim 2014; Slaby and Choudhury 2018). The natural sciences have no authority to impose normative conclusions. Such pretensions are what Hartmann (2012) calls "first order normative arguments", according to which "natural facts about the brain" can help naturalise certain political, social and economic practices or decisions. By "naturalising" them, they are seemingly "depoliticised", and therefore fall within what we call "ideology" (Rose and Abi-Rached 2013; Hartmann 2012; Malabou 2008) and usually underpin the crudest of positivisms.

The development of moral neuroeducation research may offer an important impulse for the creation of egalitarian societies. However, to make any such progress all areas of scientific research would need to overcome the various uncritical assumptions that continue to be pervasive, most notably those that underpin the dissemination of scientific findings to the general public, yet also ones that operate at the heart of research institutes, universities and laboratories (Roy 2016).

To address these challenges, analysis must be undertaken of both neurocognitive systems' epigenetic potential and their plasticity in relation to the patterns of sex/gender behavioural roles. Such studies are being carried out in other fields and are beginning to reveal promising results. One area of inspirational neuroscientific research has addressed the impact of poverty on the brain. Martha J. Farah (2005, 2010) has spent over a decade using knowledge from the neurosciences in order to develop a deeper understanding of and a way to overcome the cycle of poverty. As she and her colleagues recognise:

> These findings provide a unique opportunity for understanding how environmental factors can lead to individual differences in brain development, and for improving the programmes and policies that are designed to alleviate SES- [socioeconomic status] related disparities in mental health and academic achievement. (Hackman et al. 2010: 1)

For several years, Sebastián Lipina (2016; Lipina and Evers 2017) has also studied how childhood poverty influences cognitive and emotional development. His research has led him to claim that self-regulatory development in terms of rights, dignity, capacity and social responsibilities can have cognitive implications.

The research devoted to "the stamp of poverty" (Gabrieli and Bunge 2016) provides a way of thinking about the relationships between sex/gender and neural development. It may be argued that if research on poverty provides scientific evidence showing that social conditions leave their "stamp" on the brain, then by the same token, the construction of the subject in terms of sex/gender can also leave a cognitive trace. It may also be contended that socioeconomic status and gender interact, and further study on this interaction is called for.

By going beyond the neuromyths of the sex/gender relationship and abandoning sexist prejudices in the neurosciences, possibilities are opened up to extend knowledge about human brain development, and also to develop education, social, economic and political programmes that could enhance progress towards social equality. According to Hackman and Farah:

> The biological nature of the differences documented by cognitive neuroscience can make these differences seem all the more 'essential' and immutable. However, as already reviewed, abundant evidence from developmental neuroscience contradicts the fallacy that brain development follows a fixed, innate program and suggests specific causal pathways by which socioeconomic deprivation can affect brain function. Thus, there is little evidence to suggest differences are essential or immutable. (Hackman and Farah 2009: 71)

Interest in a proper understanding of brain differences is not only academic, but eminently economic, social and political (Heckman 2006; Sokolowski and Ansari 2018). Indeed, feminist theory states that to understand and overcome inequality a more complex framework of understanding and action is required. Inequality generally arises in circumstances where various characteristics intersect. Applying intersectionality as an approach in which several identity variables can overlap may enable a clearer understanding of both individual brains and different identity groups. Obviously, including more subjects and more variables makes neuroscientific studies more complex. Such an approach would require the large-scale collaboration between scientific research teams at an international level, and funding from state and private agencies for scientific research. Regrettably, the current state of funding for scientific research would not seem to favour to such a proposal. Instead, it is more likely that individual research teams will have their eureka moment in isolation, and therefore be able to exploit the social and economic benefits more easily.

Research on the human brain, the most complex of all brains, requires complex and ambitious projects that can meet the demands of its complexity as well as the expectations generated. By basing research endeavours on crude neuroscientific prejudices, the brain's complexity is not only overlooked, but buried further from sight.

References

Ansari, Daniel, Bert De Smedt, and Roland H. Grabner. 2012. Neuroeducation: A critical overview of an emerging field. *Neuroethics* 5 (2): 105–117.

Battro, Aantonio M., and Daniel P. Cardinali. 1996. *Más cerebro en la educación*. Buenos Aires: La Nación.

Bluhm, Robyn. 2011. New research, old problems: Methodological and ethical issues in fMRI research examining sex/gender differences in emotion processing. *Neuroethics* 6 (2): 319–330. https://doi.org/10.1007/s12152-011-9143-3.

———. 2013. Self-fulfilling prophecies: The influence of gender stereotypes on functional neuroimaging research on emotion. *Hypatia* 28 (4): 870–886.

Bluhm, Robyn, Anne Jaap Jacobson, and Heidi L. Maibon. 2012. *Neurofeminism. Issues at the intersection of feminist theory and cognitive science*, 230–245. New York: Palgrave Macmillan.

Bluhm, Robyn, Anne Jaap Jacobson, and Heidi L. Maibon. 2008. Building bridges in neuroeducation. In *The educated brain*, ed. Antonio Battro, Kurt Fischer, and Pierre Léna, 43–58. Cambridge: Cambridge University Press.

Bruer, John. 1997. Education and the brain: A bridge too far. *Educational Researcher* 26 (8): 1–13.

Bryant, Douglas J., F. Wang, Kelley Deardeuff, Emily Zoccoli, and Chang S. Nam. 2016. The neural correlates of moral thinking: A meta-analysis. *International Journal of Computational & Neural Engineering* 3: 28–39.

Busso, Daniel, and Courtney Pollack. 2015. No brain left behind: Consequences of neuroscience discourse for education. *Learning, Media and Technology* 40 (2): 168–186.

Butler, Judith. 1990. *Gender trouble*. New York: Routledge.

Chalfin, Molly C., Emily R. Murphy, and Katrina A. Karkazis. 2008. Women's sneuroethics? Why sex matters for neuroethics. *American Journal of Bioethics* 8 (1): 1–2.

Choudhury, Suparna, and Jan Slaby, eds. 2012. *Critical neuroscience: A handbook of the social and cultural contexts of neuroscience*. Chichester: Wiley- Blackwell.

Clarke, Edward H. 1874. *Sex in education (Boston)*. Cambridge: The Riverside Press.

Crenshaw, Kimberlé. 1991. Mapping the margins: Intersectionality, identity politics and violence against women. *The Stanford Law Review* 43 (6): 1241–1299.

Derks, Jeffrey, and Lydia Krabbendam. 2013. Is the brain the key to a better understanding of gender differences in the classroom? *International Journal of Gender. Science and Technology* 5 (3): 281–291.

Donaldson, Henry Herbert. 1895. *The growth of the brain: A study of the nervous system in relation to education*. Boston: Skinner Press.

Dussauge, Isabelle, and Annelis Kaiser. 2012. Neuroscience and sex/gender. *Neuroethics* 5 (3): 211–215.

Eliot, Lise. 2011. The trouble with sex differences. *Neuron* 72 (6): 895–898.

———. 2013. Single-sex education and the brain. *Sex Roles* 69 (7–8): 363–381.

Evers, Kathinka. 2015. Can we be epigenetically proactive? In *Open MIND: 13(T)*, ed. T. Metzinger and J.M. Windt. Frankfurt and Main: MIND Group.

Farah, Martha J. 2005. Neuroethics: The practical and the philosophical. *Trends in Cognitive Sciences* 9 (1): 34–40.

———. 2010. Mind, brain and education in socioeconomic context. In *The developmental interplay of mind, brain and education*, ed. M. Ferarri and L. Vuletic. Dordrecht: Springer.

Fausto-Sterling, Anne. 2000. Sexing the body. In *Gender politics and the construction of sexuality*. New York: Basic Books.

Fine, Cordelia. 2010. *Delusions of gender: How our minds, society and neurosexism create difference*. London: Icon Books.

———. 2012. Explaining, or sustaining, the status quo? The potentially self-fulfilling effects of 'hardwired' accounts of sex differences. *Neuroethics* 5 (3): 285–294.

Fumagalli, Manuela, Maurizio Vergari, Patrizio Pasqualetti, Sara Marceglia, Francesca Mameli, Roberta Ferrucci, et al. 2010a. Brain switches utilitarian behavior: Does gender make the difference? *PLoS One* 5: e8865.

Fumagalli, Manuela, Roberta Ferrucci, Francesca Mameli, Sara Marceglia, Simona Mrakic-Sposta, Stefano Zago, et al. 2010b. Gender-related differences in moral judgments. *Cognitive Processing* 11: 219–226.

Gabrieli, John D.E., and Silvia A. Bunge. 2016. The stamp of poverty. *Scientific American Mind* 28: 54–61.

Garrigan, Beverly, Anna-Lynne Adlam, and Peter E. Langdon. 2016. The neural correlates of moral decision-making: A systematic review and meta-analysis of moral evaluations and response decision judgements. *Brain and Cognition* 108: 88–97.

Geake, John. 2008. Neuromythologies in education. *Educational Research* 50 (2): 123–133.

Gilligan, Carol. 1982. *In a different voice: Psychological theory and women's development*. Cambridge, MA: Harvard University Press.

Greene, Joshua D. 2009. Dual-process morality and the personal/impersonal distinction: A reply to McGuire, Langdon, Coltheart, and Mackenzie. *Journal of Experimental Social Psychology* 45 (3): 581–584.

Grosz, Elizabeth. 1994. *Volatile bodies: Toward a corporeal feminism*. Bloomington/Indianapolis: Indiana University Press.

Gur, Ruben C., Faith Gunning-Dixon, Warren B. Bilker, and Raquel E. Gur. 2002. Sex differences in temporo-limbic and frontal brain volumes of healthy adults. *Cerebral Cortex* 12: 998–1003.

Hackman, Daniel A., and Martha J. Farah. 2009. Socioeconomic status and brain development. *Trends in Cognitive Sciences* 13: 65–73.

Hackman, Daniel A., Martha J. Farah, and Michael J. Meaney. 2010. Socioeconomic status and the brain: Mechanistic insights from human and animal research. *Nature Reviews*. 11: 651–659.

Halleck, Reuben Post. 1896. *The education of the central nervous system, 2010*. Charleston: Nabu Press.

Halpern, Diane F., Lise Eliot, Rebecca S. Bigler, Richard A. Fabes, Laura D. Hanish, Janet Hyde, Lynn S. Liben, and Carol Lynn Martin. 2011. The pseudoscience of single-sex schooling. *Science* 333 (6050): 1706–1707.

Harenski, Carla L., Olga Antonenko, Matthew Shane, and Kent Kiehl. 2008. Gender differences in neural mechanisms underlying moral sensitivity. *Social, Cognitive, and Affective Neuroscience* 3 (4): 313–321.

Hartmann, Martin. 2012. Against first nature. In *Critical theory and neurosciences*, ed. S. Choudhury and J. Slaby, 67–84. London: Wiley.

Heckman, James J. 2006. Skill formation and the economics of investing in disadvantaged children. *Science* 312 (5782): 1900–1902.

Holtzman, Geoffrey S. 2017. Neuromoral diversity: Individual, gender, and cultural differences in the ethical brain. *Frontiers in Human Neuroscience* 11: 1–4.
Howe, Julia Ward. 1874. *Sex and education: A reply to Dr. E.H. Clarke's "Sex in education"*. Edited and with an Introduction by Julia Ward Howe. Boston: Roberts Brothers.
Hyde, Janet S. 2007. New directions in the study of gender similarities and differences. *Current Directions in Psychological Science* 16 (5): 259–263.
———. 2016. Sex and cognition: Gender and cognitive functions. *Current Opinion in Neurobiology* 38: 53–56.
Jaffee, Sarah, and Janet S. Hyde. 2000. Gender differences in moral orientation: A meta- analysis. *Psychological Bulletin* 126: 703–726.
Johnson, Joy L., and Robyn Repta. 2012. Sex and gender: Beyond the binaries. In *Designing and conducting gender, sex, and health research*, ed. J.L. Oliffe and L. Greaves, 17–37. Thousand Oaks: Sage.
Jordan-Young, Rebecca M. 2010. *Brain storm: The flaws in the science of sex differences*. Cambridge, MA: Harvard University Press.
Kaiser, Annelis. 2012. Re-conceptualizing "sex" and "gender" in the human brain. *Zeitschrift fur Psychologie* 220 (2): 130–136.
Kaiser, Annelis, and Isabel Dussauge. 2015. Feminist and queer repoliticizations of the brain. EspacesTemps.net. *Travaux*, 13.04.2015.
Keller, Fox, and Evelyn. 2010. *The mirage of a space between nature and nurture*. Durkham/London: Duke University Press.
Kohlberg, Lawrence. 1976. Moral stages and moralization: The cognitive developmental approach. In *Moral development and behavior: Theory, research, and social issues*, ed. T. Lickona, 31–53. New York: Holt, Rinehart & Winston.
———. 1984. Moral stages and moralization: The cognitive developmental approach. In *Essays on moral development: Vol. 2. The psychology of moral development: The nature and validity of moral stages*, 170–205. San Francisco: Harper & Row.
Kraus, Cynthia. 2011. Critical studies of the sexed brain: A critique of what and for whom? *Neuroethics* 5 (3): 247–259.
Liben, Lynn. 2015. Probability values and human values in evaluating single-sex education. *Sex Roles* 72: 401–426.
Lipina, Sebastián J. 2016. *Pobre cerebro: los efectos de la pobreza sobre el desarrollo cognitivo y emocional, y lo que la neurociencia puede hacer para prevenirlos*. Buenos Aires: Siglo Veintiuno Editores.
Lipina, Sebastián J., and Kathinka Evers. 2017. Neuroscience of childhood poverty: Evidence of impacts and mechanisms as vehicles of dialog with ethics. *Frontiers in Psychology* 8: 61. PMC.
Lorber, Judith. 1996. Beyond the binaries: Depolarizing the categories of sex, sexuality, and gender. *Sociological Inquiry* 66 (2): 143–159.
Lüders, Ellen, Katherine L. Narr, Paul M. Thompson, David E. Rex, Lutz Jancke, Helmuth Steinmetz, and Arthur W. Toga. 2004. Gender differences in cortical complexity. *Nature Neuroscience* 7: 799–800.
Maccoby, Eleanor E., and Carol N. Jacklin. 1974. *The psychology of sex differences*. Standford: Standford University Press.
Malabou, Catherine. 2008. *What should we do with our brain?* New York: Fordham University Press.
Maney, Donna L. 2016. Perils and pitfalls of reporting sex differences. *Philosophical Transactions B. The Royal Society Publishing* 371 (1688): 371.
Pullicino, Patrick. 2014. Do genders have same brain? *Times of Malta*, 12-1-2014. Available: https://www.timesofmalta.com/articles/view/20141201/opinion/Do-genders-have-same-brain-.546355
Repo, Jemina. 2015. *The biopolitics of gender*. New York: Oxford University Press.
Reverter-Bañón, Sonia. 2016. Reflexión crítica frente al neurosexismo. *Pensamiento* 72 (273): 959–979.

———. 2017. El Neurofeminismo frente a la investigación sobre la diferencia sexual. *Daimon* 6: 95–110.
Rippon, Gina, Rebecca Jordan-Young, Anelis Kaiser, and Cordelia Fine. 2014. Recommendations for sex/gender neuroimaging research: Key principles and implications for research design, analysis, and interpretation. *Frontiers in Human Neurosciences* 8: 650.
Rippon, Gina, Rebbecca Jordan-Young, Anelis Kaiser, Daphna Joel, and Cordelia Fine. 2017. Journal of neuroscience research policy on addressing sex as a biological variable: Comments, clarifications, and elaborations. *Journal of Neuroscience Research* 95: 1357–1359.
Robertson, Diana, John Snarey, Opal Ousley, Harenski Keith, F. DuBois Bowman, Rick Gilkey, and Clinton Kilts. 2007. The neural processing of moral sensitivity to issues of justice and care. *Neuropsychologia* 45 (8): 755–766.
Rose, Nikolas, and Joelle M. Abi-Rached. 2013. *Neuro: The new brain sciences and the management of the mind*. Princeton: Princeton University Press.
Roskies, Adina. 2002. Neuroethics for the new Millenium. *Neuron* 35 (1): 21–23.
Roy, Deboleena. 2016. Neuroscience and feminist theory: A new directions essay. *Signs: Journal of Women and Culture in Society* 41 (3): 531–552.
Runnels, Vivien, Sari Tudiver, Marion Doull, and Madeleine Boscoe. 2014. The challenges of including sex/gender analysis in systematic reviews: A qualitative survey. *Systematic Reviews. BioMedCentral* 3: 33.
Russett, Cynthia E. 1989. *Sexual science: The Victorian construction of womanhood*. Cambridge, MA: Harvard University Press.
Sadker, David, and Karen Zittleman. 2009. *Still failing at fairness: How gender bias cheats girls and boys in school and what we can do about it*. New York: Scribner.
Scheele, Dirk, Nadine Striepens, Keith M. Kendrick, Christine Schwering, Janka Noelle, et al. 2014. Opposing effects of oxytocin on moral judgment in males and females. *Human Brain Mapping* 35: 6067–6076.
Schleim, Stephan. 2014. Critical neuroscience –or critical science? A perspective on the perceived normative significance of neuroscience. *Frontiers in the Human Neuroscience* 8 (336).
Schmitz, Sigrid. 2012. The neurotechnological cerebral subject: Persistence of implicit and explicit gender norms in a network of change. *Neuroethics* 5 (3): 261–274.
Slaby, Jan, and Suparna Choudhury. 2018. Proposal for a critical neuroscience. In *The Palgrave handbook of biology and society*, ed. M. En Meloni, J. Cromby, D. Fitzgerald, and S. Lloyd. London: Palgrave Macmillan.
Sokolowski, Moriah H., and Daniel Ansari. 2018. Understanding the effects of education through the lens of biology. *Science of Learning* 17: 1–10.
Spelman, Elizabeth V. 2007. Managing Ignorance. In *Race and epistemologies of ignorance*, ed. S. Sullivan and N. Tuana, 119–134. New York: Suny Press.
Spivak, Gayatri Chakravorty. 1993. *Outside in the teaching machine*. New York/London: Routledge.
Theodoridou, Zoe, and Lazaros Triarhou. 2009. Fin-de-Siecle advances in neuroeducation: Henry Herbert Donaldson and Reuben Post Halleck. *Mind, Brain and Education* 3 (2): 119–129.
Tuana, Nancy. 2004. Coming to understand: Orgasm and the epistemology of ignorance. *Hypatia: A Journal of Feminist Philosophy* 19 (1): 194–232.
———. 2006. The speculum of ignorance: The women's health movement and epistemologies of ignorance. *Hypatia: A Journal of Feminist Philosophy* 21 (3): 1–19.
Unger, Rhoda K. 1979. Toward a redefinition of sex and gender. *American Psychologist* 34 (11): 1085–1094.
Vidal, Catherine. 2012. The sexed brain: Between science and ideology. *Neuroethics* 5 (3): 295–303.
Walker, Lawrence J. 2006. Gender and morality. In *Handbook of moral development*, ed. M. Killen and J.G. Smetana, 93–115. Mahwah: Erlbaum.

Chapter 12
Neuroleadership: Diversity as a Moral Value in Organisations

Maria Medina-Vicent

12.1 Introduction

The term neuroleadership (Rock 2009) refers to the application of neuroscience to the field of business leadership. The primary aim of this dialogue between neuroscience and business is to improve the management, productivity and leadership of organisations. Underpinning this exchange is the fact that a leader can develop skills that both enhance efficiency, yet also steer themselves away from distracting practices or behaviour that would hinder efficiency.

The importance of these discoveries for business organisations has become increasingly apparent since the emergence of "brain gyms", Brain Management Training seminars and even the vogue for mindfulness. The development of such training methods in conjunction with new findings on how the brain works, provided by technologies such as magnetic resonance and magnetoencephalography, offer a vast spectrum of potential for organisational management to successfully use each manager's cerebral capacities in the most appropriate manner. Some manuals also provide practical tools for transforming management based on the neurosciences (Rock and Page 2009; Herreros 2012), such as the *SCARF* model (*Status, Certainty, Autonomy, Relationships and Fairness*) advocated by David Rock, director of *The NeuroLeadership Institute*. Neuroleadership studies the mental processes of successful managers and business leaders to provide insights into how a knowledge of brain functions that can lead to greater business productivity, which is why

This study is part of the Scientific Research and Technological Development Project: "Neuroeducación Moral para las Éticas Aplicadas" [FFI2016-76753-C2-2-P] funded by the *Ministerio de Economía y Competitividad.*

M. Medina-Vicent (✉)
Universitat Jaume I, Castellón de la Plana, Spain
e-mail: medinam@uji.es

P. Calvo, J. Gracia-Calandín (eds.), *Moral Neuroeducation for a Democratic and Pluralistic Society*, https://doi.org/10.1007/978-3-030-22562-9_12

the core issues that neuroleadership studies focus on is decision-making, as well as memory, attention, motivation and emotional intelligence (Ghadiri et al. 2012).

Based on these issues, team-building and a combination of different leadership styles has become one of the central premises of the convergence between neuroscience and leadership (Ringleb and Rock 2009; Gordon 2013; Smith et al. 2016; Rock and Grant 2016). The value of diversity—which has been acknowledged in economic theory for decades—is once again being revived in the business world, on the basis that it contributes a greater productivity and efficiency, a view that neuroscientific discourse now lends weight to. Taking into account neuroleadership's interest in the management of companies' diversity, this chapter considers the potential that could be provided by forging synergies between neuroleadership and business ethics in order to enhance the value of diversity. In other words, from an ethical perspective, we must consider whether the breakthroughs in neuroleadership as a discipline can foster diverse work teams, or whether the quest for the goal of diversity is in fact merely a neoliberal discourse on individualism, self-management and self-discipline.

Over the course of this chapter I examine the concept of neuroleadership and establish a distinction between leadership and management, which provides a basis to understand the importance of shifting from an individual view of management processes to a collective perspective related to business management. Attention is then turned to the meaning of individualism within the discourse of neuroleadership, and this is examined in a framework of vigorous neoliberalism so as to foreground the necessity of establishing an ethical form of management. Finally, diversity as a potential convergence between neuroleadership and ethical business management is examined (Cortina 1994a, 2013; Conill 2003b; García-Marzá 2004; González-Esteban 2016).

12.2 An Approach to the Concept of "Neuroleadership"

Although the literature on leadership is wide-ranging and prolific, there is no agreement as how to successfully undertake leadership (Rock 2010). As a result, the application of neuroscience to the study of leadership has become one of the most rapidly expanding areas of study in recent years. As a concept, neuroleadership was coined by the author David Rock, director of the *Neuroleadership Institute*, in 2006 in the journal *Strategy + Business* (Rock and Schwartz 2006, 2010), and he later developed the idea in greater depth in his study entitled *Your Brain at Work* (Rock 2009). The term refers to the application of neuroscientific findings to the field of leadership. It therefore introduces knowledge provided by technologies including magnetic resonance imaging to the areas of leadership development, consulting and coaching. Rock's principal theses (Rock 2010), based on the results of neuroscientific experiments applied to the study of leadership, provide a point of departure to develop techniques and practices to improve management practices.

Firstly, there is the "Aha! moment", the instant of successfully solving a problem. John Kounios and Mark Beeman (2009) argue that those people who are more creative and tend to distance themselves from linear reasoning processes may experience more of these moments, and therefore become more decisive leaders. Closely related to this issue is neuroleadership's main aim: ensuring good decision-making, meaning decisions that increase a company's efficiency. When we feel threatened and anxiety levels are high, we may overlook crucial information when making decisions, and we may make more mistakes because we are distracted. According to neuroscience, being able to observe how our brain functions enables us to regulate our emotions in an "efficient" way. For example, studies of social pain show that the brain deals with social pain in the same way as it deals with physical pain (Eisenberger 2012). Therefore, the issue is one of being able to identify situations that create this pain and learning to "control" our brain's reaction to them. As we shall see, it is a question of modifying the brain rather than addressing social circumstances.

To encourage these "Aha! moments", and to exercise an ability to make good or efficient decisions, neuroleadership studies advise that the brain's general chatter be silenced, which entails minimising anxiety and reducing neuronal activity. The aim is to relax and turns one's focus within through *inward looking*, what is *not effortful* (doing something without trying), and *mindfulness* (Kabat-Zinn 1994, 2005). Such practices are based on the thesis of a regulation of the emotions, which is related to what Lieberman (2009) has called the "brain's braking system", a system that causes emotions to become weaker when it is activated.

As with other areas related to cerebral neuroplasticity (Wexler 2011), the braking system can be trained. The task for leaders is to learn to activate it when they feel very strong emotions which prevent them from making good decisions. Some neuroleadership programmes, like the one developed by *Zero Point Leadership*, use biofeedback techniques such as the *emWave2* mobile technology developed by HeartMath® to help leaders modify their reaction to stress, and to learn to activate the braking system. Situations that create stress are therefore individually and internally "neutralised". Responsibility lies with the individual; there is no overall or social consideration of the causes that create this stress. Related to this is the reappraisal or re-evaluation technique (Ochsner et al. 2002, 2004; Ochsner and Gross 2005), which is based on the ability to interpret things differently, or to change a situation's meaning. Recontextualising and re-evaluating certain issues that are problematic in our daily working routine enables us to change our emotional responses to them.

As can be seen, most of the strategies and practices in neuroleadership revolve around the question of decision-making and regulating one's emotions. In other words, this discipline studies how we can train our brain to avoid making decisions that may be harmful to an organisation. According to this perspective, we will make better decisions if we control the strong emotions that arise in stressful situations. However, neuroleadership does not address the conditions that create this emotion, i.e. the injustices and inequalities that lie behind the strong emotions felt by leaders and other team members; in contrast business ethics does, as is discussed below.

To a large extent, this can mean losing sight of an important issue, which is that a sense of threat or fear can help us identify situations of injustice and take actions that are morally good. For example, some situations that give rise to sadness can lead to both individual and collective action. If leaders suppress their capacity to feel sad, they may not be able to improve the organisation's working dynamics. Furthermore, the said ability to identify situations of injustice should not be solely the responsibility of leaders, but also the institution itself. Therefore, having identified this logic, it is therefore necessary to consider a transition between the individual dimension outlined by neuroleadership, and the institutional dimension of business management. So, in order to develop this discussion further consideration must first be given to the distinction between concepts of leadership and management. Establishing the differences between them provides a foundation for a deeper understanding of the implications of neuroleadership for business management.

12.3 Leadership vs Management: Establishing the Difference

When the practice of leadership is transformed the change influences a company's management. As a result, the changes that subsequently take place in management vary depending on a business leader's approach to neuroleadership.[1] Let us therefore begin by defining leadership, which is a widely studied concept.[2] Solid benchmarks for reflection on this issue are provided by the *Handbook of Leadership* (Bass 1990) and the *Encyclopaedia of Leadership* (Goethals et al. 2004), which cover a range of issues related to the definition of this concept and its problems.

In addition to specific definitions of leadership models, as has been said this chapter seeks to ascertain the basic differences between leadership and management, so as to better understand the implications of the influence of the discourse of neuroleadership on business management. The book *Leading Organizations: Perspectives for a New Era* (1998), edited by Gill Robinson Hickman, includes contributions from leading management and leadership theorists. It also contains Joseph C. Rost's[3] valuable definition of what leadership is and how it differs from

[1] Within the field of the neurosciences the concept of Neuromanagement is also used, but it is used almost indistinguishably from neuroleadership (Braidot 2013).

[2] To illustrate the wide variety of approaches to the study of leadership, some of the models that have had the most influence on leadership theory in recent decades are listed below: *Transactional Leadership* (Bass et al. 1996), *Transformational Leadership* (Bass and Riggio 2006), *Charismatic Leadership* (Conger and Kanungo 1998), *Ethical Leadership* (Brown et al. 2005; Brown and Treviño 2006), *Eco-Leadership* (Western 2012, 2013), *Narcissistic Leadership* (Rosenthal and Pittinsky 2006), *Global Leadership* (Adler et al. 2000; Adler 1997), and *Spiritual Leadership* (Ngunjiri 2010).

[3] Joseph C. Rost is the author of *Leadership for the Twenty-first Century* (1991), in which he set out the challenges for leadership practice in the global context of the twenty-first century. He has also undertaken research at the University of San Diego, on various subjects related to leadership and ethics (Rost 1993, 1995, 1997; Rost and Barker 2000).

the concept of management. Given its conceptual clarity and accuracy, both of which are central to the modest task undertaken in this chapter, his definition stands out from the others; despite their being also very interesting, the latter lie beyond the scope of this study. According to Joseph C. Rost (1998), the four main differences between leadership and management are: a relationship of influence rather than one of managerial authority; leadership establishes a leaders-followers dynamic, while management a managers-subordinates one; leaders seek to make real changes rather than solely develop the production and sale of goods and services; and, finally, leaders are concerned with shared objectives rather than the managerial role of coordinating activities. We will now discuss these four differences.

Firstly, we will consider the distinctive types of relationship between a leader, on the one hand, and a manager, on the other, and their team members. Leadership is an influence-based relationship, while management is an authority-based relationship. The use of coercion and a sense of the impact one has on people are crucial in both cases, albeit in different ways. When the aim is to influence and inspire others' behaviour, coercion must be set aside, whereas when the aim is to establish a sense of authority in the group, the use of coercion may become part of the manager's daily behaviour. Therefore, the process of influence involved in leadership is multidirectional, as it depends on the attitude and commitment of the leader's followers, but in management practice it is unidirectional, as it solely seeks to coordinate actions and make them effective. Although Rost acknowledges that the managerial process may be more democratic, the basic characteristics of the relationship it entails are those outlined above. This means that the premise of "managing" emotions in neuroleadership has a direct impact on processes linked to a leader's influence.

A second difference between the leader and the manager relates to teamwork. As has been discussed with regard to leaders the team is understood as comprising followers, whereas in the case of a manager's team, it is made up of subordinates. An important distinction is made by Rost (1998: 112): "Leaders need to be managers to be leaders. Followers need to be subordinates to be followers". In other words, managers can be leaders, but if so, they are engaged in a different relationship to what is understood by management. The same applies to subordinates, who can also be followers, but again in this case a different relationship to that of management is invoked.

An important idea emerges from this reflection on leadership and management: people in positions of importance, involving responsibility and management, do not automatically become leaders. Instead, in order to achieve the transition from managers to leaders they need to build a different relationship with their work teams, one based on an intention to conduct real change and share a common vision. Occupying a position of authority does not make a person a leader, but rather a manager, since this is the definition of a manager: a person who occupies a position of authority. According to Rost, the leader-manager, follower-subordinate set is not interchangeable under any circumstances. From this perspective, it can be seen that a leader will have much more influence on their team, given that they are a source of inspiration to others. It is therefore easier for a team of followers to incorporate the values of leadership that motivate and inspire, rather than the values dictated by a company's hierarchical authority. However, this question also raises an issue that cannot be ignored, which is that the business world is a sphere in which social hierarchies are reproduced, and which contains tensions that must be manageable. Not everything depends on the role of the individual. The prevailing social context is another factor, and neuroleadership seemingly devotes little consideration to this issue.

Thirdly, there is the question of the overarching goals of leadership and management. While the main aim of the former is to create real change, the latter focuses on producing/selling goods and/or services. In other words, leaders and followers join forces to effect some sort of change either in the organisation or in another area, which signfies that their practice means something other than merely engaging in their profession; meanwhile, managers and their subordinates solely aim to manage the production and sale of their organisation's products or goods in an increasingly effective way, meaning that any changes that take place will always be aimed at these efficiency-related goals.

Finally, leadership practice involves a common goal, while for management there is a group of people who merely coordinate their actions to carry out activities. There are no shared meanings, values or goals in Rost's definition of management, although they can be created as tasks are carried out. In the event this happens, Rost says that this form of management becomes a leadership practice.

In short, according to Rost, leaders must be able to influence and inspire others to enrich a shared culture, and therefore they work with values; while a manager's role is limited to management. However, the distinction made by Rost should not be taken literally, since the practice of leadership affects management, and management is in turn determined by leadership. In other words, the ability to generate values is not solely the preserve of leadership, but is also an outcome of business management, which is ultimately responsible for making a company function, regardless of any role played by an individual company leader. This means that, to a certain extent, there will always be a need to manage the individual dimension of leadership within a structural and organisational sense of the company as institution. It is also necessary to consider the values underlying the discourse of neuroleadership, since these will in turn influence the formation of a company's shared culture. In the following sections attention is devoted to the principal logic that arises from neuroleadership's emphasis on individualism, and this is compared with the need to articulate an ethical mode of business management.

12.4 Neuroleadership: A Disciplinary Technique?

As observed in the previous sections, an individual cannot be a leader without first occupying the role of manager in a company. However, becoming a leader involves a great deal more, as it means influencing followers and guiding them towards a common goal. To achieve this, leaders have to situate themselves within their immediate surroundings, and address their group's needs, and the necessary dynamics to develop their practice, as well as following the directions imposed by their company's structure and management. Nonetheless, closer scrutiny of the assumptions underpinning neuroleadership, reveals a clear emphasis on personal individuality, and leaders are to some extent isolated from their surroundings and the people that make up their team. In other words, they try to calm strong emotions, which may be the result of unjust situations, whereby their capacity for self-control exerts itself with greater force and that of self-management plays a much more prominent role.

To a certain degree, the arguments set out in studies on neuroleadership and the practices and techniques derived from them encourage a form of leadership that does not stop to consider situations that give rise to pain or threats. In many cases, these situations are those that require the most consideration, because they are the ones that arise from injustice. As a result, the language of neuroleadership commodifies the emotions, i.e. the emotions are placed at the service of the logic of the market, which thereby leads to emotional capitalism (Illouz 2007, 2010). Market logics such as this are integrated into the business world through the use of psychology to ensure individuals adhere to the latest commercial and labour dynamics. In this regard, neuroleadership could act as a disciplinary tool for the self-regulation for leaders that operate in accordance with a mercantile logic; through self-regulation based on the control of their emotions, the leader is distanced from any in-depth consideration of situations that arouse strong emotions, and this also distances them from the structural sense of business management.

The neoliberal era (Boltanski and Chiapello 2002) requires bureaucratisation systems that are much more sophisticated than those of previous periods. These are systems that guarantee both public management and privatisation, as well as management of both one's self and business (Foucault 2007). It is a bureaucracy that focuses on controlling the population to ensure the divulgation of the neoliberal spirit and commercial logic. As a result, each individual exercises self-control over their daily actions, which ensures that they are completely prepared for changes and subsequent market crises (Alonso and Fernández Rodríguez 2013a), and their leaders are prepared for subsequent instrumental requirements. Furthermore, an individual's incorporation of a logic of self-control to a large extent acquires meaning through the social model of the manager, which is disseminated and functions as a mirror in which to see oneself (Alonso and Fernández Rodríguez 2013b) and that in turn establishes autonomously isolated subjects (Laval and Dardot 2013).

Thereby, within today's voracious capitalist society, a cultural operation is thereby conducted according to which solutions to social problems lie within individuals (Papalini 2006: 42), and the role of the state and organisations in solving them is denied. Individuals become responsible for autonomously regulating their emotions (rage, sadness, impotence, etc.) in order to remain in the market and survive (Muñoz-Rodríguez and Santos-Ortega 2017). The same logic also seems to emerge, albeit in a very diluted form, within the discourse of neuroleadership. As a result, from a critical standpoint, it may be argued that the discourse of neuroleadership risks perpetuating an individualistic and biased view of the problems that may arise within organisations and therefore in society; problems that must be addressed in terms of a dialogue between all the affected parties. Therefore, ethical management is generally required to address problems that may arise within a company. Through this ethic the discourse of individualism and isolation inherent in neoliberalism is dismantled and replaced with an overarching consideration of the situations of injustice that occur in the company, and that result in the aforementioned "strong emotions". It is at this point that a proposal for business ethics that addresses the needs of the groups affected by a particular activity comes into play, and this must acknowledge institutions' collective responsibility for their actions and decisions.

12.5 From Leadership to Ethical Management

Despite the neoliberal current within the discourse of neuroleadership, in general terms it seeks to appeal to an organisation's human dimension rather than any functional concept of it (Turnbull 2015; Rock and Grant 2016; Cox et al. 2017). However, Rock, who could be considered the father of this discipline, acknowledges that we must use the language of economic competitiveness when incorporating the breakthroughs of neuroscience into business management. Therefore, to a great extent the difference between the maximising and individualistic language used by neuroleadership and the moral language of ethics is an outcome of the search for greater success in business management, which continues to be closely linked to the logic of *homo economicus*. As a consequence, the discourse of neuroleadership contributes, to a certain degree, to the perpetuation of an individualistic concept of leadership that serves only to legitimise the new spirit of capitalism (Boltanski and Chiapello 2002), while also endangering neuroleadership's potential contribution to organisational management. As is apparent, this is a wholly different language from that of ethics, which is informed by moral values. Although, on occasions the discourse of neuroleadership does appeal to an organisation's human dimension, closer scrutiny reveals it cannot transform contemporary economic rationality without addressing the structural dimension that sustains it.

In contrast, the proposal for business ethics developed by the research group dedicated to the study of applied ethics and democracy at the Universidad de Valencia and that dedicated to research on practical ethics and democracy at the Universitat Jaume I (Cortina 1994b, 2013; Conill 2004; García-Marzá 2004; González-Esteban 2007) enables us to reconsider the rationality of neuroleadership by undertaking an ethical reflection on a company's responsibility for the actions and decisions taken with regard to its environment and human groups, which could be achieved through a dialogue between the interest groups affected by the company's work. Only in this way can a business ethos, and with it the functionalist economic rationality, undergo transformation. From an ethical perspective, this entails transforming the foundations of the current economic rationality and turning away from the model of self-interested rationality (Cortina 1994b: 65), which is inextricably related to the predominant logic of self-control in the discourse on neuroleadership. What is proposed is a cognitivist ethic, in which the moral dimension does not refer directly to rules, but instead the claim of moral validity underpinning the rules. Moral norms or values operate because they are valid for all those interested in them, and not because they prevail within society, and it is in this way that their existence is morally justified.

It is human beings' shared ability to use language to communicate that shows us whether a rule is correct, righteous or just, and this is achieved through intersubjective agreements based on the common interests involved in each case (García-Marzá 1992: 28–41). There are therefore procedural ethics, in which "the normative content of this epistemic language game is transmitted only by a rule of argumentation to the selection of norms of action, which together with their moral validity claim provide the input into practical discourses" (Habermas 1998: 45). In other words, a proposal for a business ethics based on dialogue supersedes the individual and isolated vision of business management promoted by the discourses of neuroleadership, and this would offer a vision of the organisation that addresses the interests of all groups affected by the its activities. Thereby, a transition is effected from the more individual practice of leadership to a consideration of the structure of business management. Within this perspective, individuals are obliged to come out of themselves in order to consider and listen to the needs of others, and this also occurs within the organisation, which contrasts to the logic of neuroleadership discussed above.

While neuroleadership offers a vision of the problems each individual may experience in the business environment, a dialogical business ethics reminds us that the roots of these problems are organisational, and as such must be addressed in coordination with all the groups affected (Cortina 2003; Conill 2003a; García-Marzá 2012). Moral language and feelings are crucial in this deliberation (García-Marzá 2005), as they alone offer a transition from extreme individual responsibility, a trait of the neoliberal framework, to a new ethical framework in which business organisations acknowledge their moral responsibility for injustices committed within them. Accordingly, we must remember that the organisation is a social space defined by power tensions and inequalities based on gender and other issues (Medina-Vicent 2017a). Issues such as these cannot be resolved individually. They require a global perspective and the tools that ethical business management provides.

12.6 Diversity as a Moral Value in Organisations

Drawing on the issues addressed so far, attention is now turned to the concept of diversity in order to ascertain the differences between the predominant instrumental and individualistic concept of neuroleadership, and the moral and structural concept of business ethics. Diversity is one of the central moral values in organisational management, and the exercise of this value to some extent overcomes the concept of the individual encouraged by the discourse of neuroleadership. The value ascribed to diversity has increased in the business world because it is considered to contribute greater productivity, but it may be asked what assumptions underpin the incorporation of this value into productivity management.

Before continuing, let us clarify what is meant by diversity, since it is another term on which there is no clear consensus (Susaeta and Navas 2006). Kossek and Lobel (1996) consider diversity solely in terms of differences in race, gender and ethnicity. Meanwhile, other authors such as Carnevale and Stone (1995), Norton and Fox (1997), Triandis (2003) and Carr (2008) include other factors, such as age, nation of origin, religion, sexual orientation, disability, values, ethnic culture, language, lifestyle, beliefs, physical appearance and economic status. On the other hand, authors such as Hon and Brunner (2000), emphasise diversity in terms of issues related to the various types of discrimination different human groups suffer as a result of the constraints they face.

The gradual rise of the importance of diversity in companies is related to the appearance of and increase in diversity policies in companies, which in the United States took place in the late 1980s. At that time, an academic debate was taking place about the effectiveness of Affirmative Action, a series of legal measures implemented in the 1960s and 1970s to guarantee minority groups equal opportunities in access to employment (Chinchilla and Cruz 2011: 48). Some authors began highlighting the need to go beyond the legal sense of these actions, or in other words to understand diversity not only as a way of complying with legal requirements, but also as a source of efficiency and competitive advantage for companies (Thomas 1990; Cox and Blake 1991).

Bearing in mind these issues, the definition of diversity addressed here refers to a scenario in which the differences between individuals within an organisation are taken into account, and the resulting heterogeneous work groups acquire a raison d'être in a global context. It is in this global context, that injustices, misunderstandings and disparities occur with greater ease (Basset-Jones 2005), although it is generally argued that it also generates more interesting and creative synergies for business management (Rock et al. 2016). It is this issue that causes the most concern, as a company's diverse work groups are also more likely to reproduce the dynamics of inequality that exist in society. For this reason, it is relevant to consider how neuroleadership understands diversity, and thereby one can discern whether this contributes to neuroleadership's neoliberal tendency, or whether it can provide a basis for cooperation between neuroleadership studies and business ethics.

By way of an example, let us return to the *Neuroleadership Institute*, and in particular Smith and Rock's article "Approaching Diversity with the Brain in Mind" (2018), which focuses on how the brain's unconscious biases hinder diversity and inclusion within a company. The authors point out that in order to counteract the unconscious biases that hinder the creation of an inclusive and efficient space, it is necessary to go beyond education, since education does not transform an agents' behaviour, but instead merely makes them aware of the existence of these biases:

> The most important finding here is a broad one: For decades, psychologists and political scientists have looked for evidence that educating people about certain issues causes them to act differently. Again, and again, on public health issues such as antismoking and anti-drug campaigns, they've come up empty. People smoke despite knowing the risks of tobacco use. [...]

> From a neuroscientific perspective, none of this should come as a surprise. Unconscious biases occur as a result of brain processes that aren't consciously accessible, so it's only natural that talking and thinking about them doesn't really change anything. Of course, education and awareness efforts can get us to think more about diversity and inclusion, and can spark interest in addressing those goals, but they aren't enough to change the unconscious processes that lead to bias in the first place.

An idea seemingly set out in this article is that: education makes people aware, but it does not change their behaviour. The power of education to exercise individuals' critical capacity is thereby devalued. Therefore, on reading arguments of this type one might expect that neuroleadership would provide the tools or solutions that are not based on dialogue, language or deliberation, but instead on "brain" or "neurological" mechanisms, given that according to neuroleadership, "it's only natural that talking and thinking about them does not really change anything" (Smith and Rock 2018). However, one simple piece of advice is given: train and teach teams to identify biases when they occur and to verbalise those situations. At first glance, it appears that there is no significant difference. In addition, a series of measures and programmes to encourage changes of behaviour within a company's groups are also proposed. The training proposed includes nine self-guided webinars, a three-hour workshop, an interactive experience with other members of the company, and the construction of inclusive networks.[4] Evidently there is a marked ambivalence towards the power of education, awareness and dialogue, which was seemingly rejected at the outset, only for measures based on those very components to be subsequently proposed.

If we restrict attention to the general logic of neuroleadership as discussed above, it is apparent that the answers provided by this discipline in terms of encouraging diversity and inclusion in companies are not very different from approaches that business ethics has been offering for some years. While neuroleadership discusses talks, training and workshops, business ethics advocates applying an integrated system of ethical management, including instruments such as ethical codes, ethics committees, social initiatives and ethical audits (García-Marzá 2004). One of the common principles linking the issues raised by the two disciplines is language and dialogue, whereby there may be the possibility of a convergence between these conceptual approaches. While, the seeming persistence of a more individualistic concept within the discourses of neuroleadership must not be overlooked, and that an understanding of business management as an institutional issue based on a constant dialogue between all the groups affected has to be foregrounded, (González-Esteban 2002, 2007, 2012), neuroleadership may nonetheless be able to make a valuable contribution to business diversity and moral value within organisations.

To proceed, attention must be paid to how a discipline like neuroleadership sustains a focus on managing individual stress, but we can also note how a change can occur when neuroleadership comes into contact with ethics, or in other words a convergence can be achieved between individuals' experiences of stress and the causes of it. If we consider diversity to be a moral value that must be integrated into

---

[4]The full case study is available on the *Neuroleadership Institute* website: https://neuroleadership.com/di-case-studies-f500-tech-company/

organisations along with that of equality (Medina-Vicent and Reverter-Bañón 2016; Medina-Vicent 2017b), we must therefore acknowledge that this value must not only be "taught" to groups, but also become one of the foundations for business culture and decision-making, because it is a question of social justice. And it is not only leadership that is at stake, but also business management, which means the organisation's structure, management and institutional decision-making. In other words, the value of diversity must be integrated into the organisational structure itself, by means of measures that enable organisational management to make decisions based on this value.

In short, from an ethical perspective, a consideration is required as to whether the breakthroughs of neuroleadership foster genuinely diverse work teams, or whether its aim of achieving diversity merely conceals the individualistic logics that reproduce the instrumental principles of neoliberalism. As a field of study, it is too early to tell with any certainty what effects neuroleadership may have on business management. However, the neoliberal current within its discourse, as identified over the course of this chapter, means that close attention must be paid to its evolution over the coming years, although this does not preclude its possible convergence with business ethics.

12.7 Conclusion

The significance of the discourse on neuroleadership is tied to the wider context of the current neoliberal era (Boltanski and Chiapello 2002), and it has become part of the contemporary discourse on management (Fernández Rodríguez and Medina-Vicent 2017), thereby influencing business management, and promoting a sense of management in which the self is viewed as a business (Foucault 2007). Today's neuroleadership programmes, which aim to achieve "effective" leadership for organisations, directly associate this effectiveness with a total control of the emotions. An individualistic and instrumental vision of the leadership process is thereby fostered, one that can even silence injustices that occur in the workplace through self-accountability of the individual and the depoliticization of the causes of those problems. If we curb our emotions, we are preventing any transition towards a critical reflection on the conditions of inequality that created the situation in question.

It is therefore crucially important to address a company's structures and organisational management, as well as the voices of all those affected when solving problems or making decisions. For this reason, a critical reflection on neuroscientific breakthroughs and how they are applied in the organisational field is necessary (Cortina 2011; García-Marzá 2016), as is also a consideration of how business management can be undertaken from an ethical standpoint. Although, the neoliberal current within neuroleadership discourse is all too apparent, it is important to be open to breakthroughs in organisational neuroscience and neuroethics (González-Esteban 2016), since an ethical and dialogical reflection on neuroscience can be achieved, as has been shown in other disciplines (Pallarés-Domínguez 2015).

In short, when we talk about neuroleadership, are we talking about a new discourse of neoliberal power? Or can we forge a dialogue between it and ethics? It appears to be too early to provide a categorical answer to these questions. Yet, what is clear is that a business organisation is a social area made up of power tensions and inequalities, and these issues cannot be resolved individually. Instead, they require both a global perspective and the tools that ethical management provides, tools that distance us from the instrumental and bring us closer to reciprocity (Calvo 2018).

References

Adler, Nancy J. 1997. Global Leadership: Women Leaders. *Management International Review* 37 (1): 171–196. https://doi.org/10.1007/978-3-322-90987-9_10.

Adler, Nancy J., Laura W. Brody, and Joyce S. Osland. 2000. The Women's Global Leadership Forum: Enhancing One Company's Global Leadership Capability. *Human Resource Management* 39 (2–3): 209–225. https://doi.org/10.1002/1099-050X(200022/23)39:2/3<209::AID-HRM10>3.0.CO;2-Y.

Alonso, Luis Enrique, and Carlos Jesús Fernández Rodríguez. 2013a. Debemos aplacar a los mercados: el espacio del sacrificio en la crisis financiera actual. *Vínculos de Historia* 2: 97–119.

———. 2013b. *Los discursos del presente. Un análisis de los imaginarios sociales contemporáneos*. Madrid: Siglo XXI.

Bass, Bernard M. 1990. *Handbook of Leadership. Theory, Research, and Managerial Applications*. New York: The Free Press.

Bass, Bernard M., and Donald E. Riggio. 2006. *Transformational Leadership*. Mahwah: Lawrence Erlbaum Associates.

Bass, Bernard M., Bruce J. Avolio, and Leanne Atwater. 1996. The Transformational and Transactional Leadership of Men and Women. *Applied Psychology* 45 (1): 5–34. https://doi.org/10.1111/j.1464-0597.1996.tb00847.x.

Basset-Jones, Nigel. 2005. The Paradox of Diversity Management, Creativity and Innovation. *Creativity and Innovation Management* 14 (2): 169–175.

Boltanski, Luc, and Ève Chiapello. 2002. *El nuevo espíritu del capitalismo*. Madrid: Akal.

Braidot, Néstor P. 2013. *Neuromanagement. Cómo utilizar a pleno el cerebro en la conducción exitosa de las organizaciones*. Barcelona: Granica.

Brown, Michael E., and Linda K. Treviño. 2006. Ethical Leadership: A Review and Future Directions. *The Leadership Quarterly* 17 (6): 595–616. https://doi.org/10.1016/j.leaqua.2006.10.004.

Brown, Michael E., Linda K. Treviño, and David A. Harrison. 2005. Ethical Leadership: A Social Learning Perspective for Construct Development and Testing. *Organizational Behavior and Human Decision Processes* 97 (2): 117–134. https://doi.org/10.1016/j.obhdp.2005.03.002.

Calvo, Patrici. 2018. *The Cordial Economy – Ethics, Recognition and Reciprocity*. Cham: Springer.

Carnevale, Anthony Patrick, and Susan Carol Stone. 1995. *The American Mosaic: An In-Depth Report on the Future of Diversity at Work*. New York: McGraw-Hill.

Carr, Clay. 2008. Diversity and Performance: A Shotgun Marriage? *Performance Improvement Quarterly* 6 (4): 115–126. https://doi.org/10.1111/j.1937-8327.1993.tb00608.x.

Chinchilla, Nuria, and Hugo Cruz. 2011. Diversidad y paradigmas de empresa: un nuevo enfoque. *Revista Empresa y Humanismo* 14 (1): 47–79.

Conger, Jay, and Rabindra Kanungo. 1998. *Charismatic Leadership in Organizations*. London: Sage.

Conill, Jesús. 2003a. Economía ética en la era de la información. In *Construir confianza. Ética de la empresa en la sociedad de la información y las comunicaciones*, ed. Adela Cortina, 75–96. Madrid: Trotta.

———. 2003b. El sentido ético de la economía en tiempos de globalización. *Daimon. Revista Internacional de Filosofía* 29: 9–15.
———. 2004. *Horizontes de economía ética. Aristóteles, Adam Smith, Amartya Sen.* Madrid: Tecnos.
Cortina, Adela. 1994a. *Ética de la empresa: claves para una nueva cultura empresarial.* Madrid: Trotta.
———. 1994b. *La ética de la sociedad civil.* Barcelona: Anaya.
———. 2003. *Construir confianza: ética de la empresa en la sociedad de la información y las comunicaciones.* Madrid: Trotta.
———. 2011. *Neuroética y Neuropolítica. Sugerencias para la educación moral.* Madrid: Tecnos.
———. 2013. *¿Para qué sirve realmente la ética?* Barcelona: Paidós.
Cox, Taylor, and Stacey Blake. 1991. Managing Cultural Diversity: Implications for Organizational Competitiveness. *Academy of Management Executive* 5 (3): 45–56.
Cox, Christine, Josh Davis, David Rock, Camille Inge, Heidi Grant, Kamila Sip, Jacqui Grey, and Lisa Rock. 2017. The Science of Inclusion. *Neuroleaderhsip Journal.* https://neuroleadership.com/portfolio-items/the-science-of-inclusion/ Accessed 13 Sept 2018.
Eisenberger, Naomi I. 2012. The Neural Bases of Social Pain: Evidence for Shared Representations with Physical Pain. *Psychosomatic Medicine* 74 (2): 126–135. https://doi.org/10.1097/PSY.0b013e3182464dd1.
Fernández Rodríguez, Carlos Jesús, and Maria Medina-Vicent. 2017. Los nuevos discursos del management: difusión, impactos y resistencias. *Recerca. Revista de pensament i anàlisi* 20: 7–14. https://doi.org/10.6035/Recerca.2017.20.1.
Foucault, Michel. 2007. *Nacimiento de la biopolítica.* Buenos Aires: Fondo de Cultura Económica.
García-Marzá, Domingo. 1992. *Ética de la justicia: J. Habermas y la ética discursiva.* Madrid: Tecnos.
———. 2004. *Ética empresarial: del diálogo a la confianza.* Madrid: Trotta.
———. 2005. Sentimientos virtuosos. El papel de los sentimientos en la vida moral. *Diálogo Filosófico* 62: 241–256.
———. 2012. Business Ethics as Applied Ethics. A Discourse Ethics Approach. *Ramon Llull Journal of Applied Ethics* 3: 99–114. https://doi.org/10.4103/2013-8393.107302.
———. 2016. Neuroética aplicada: las consecuencias prácticas del neuropositivismo. *Pensamiento. Revista de investigación e Información filosófica* 72 (273): 881–900. https://doi.org/10.14422/pen.v72.i273.y2016.007.
Ghadiri, Andreas, Argang Habermacher, and Theo Peters. 2012. *Neuroleadership: A Journey Through the Brain for Business Leaders.* New York: Springer.
Goethals, George R., Georgia J. Sorenson, and James MacGregor. 2004. *Encyclopedia of Leadership.* Thousand Oaks: Sage.
González-Esteban, Elsa. 2002. Defining a Post-conventional Corporate Moral Responsibility. *Journal of Business Ethics* 39: 101–108. https://doi.org/10.1023/A:1016388102599.
———. 2007. La teoría de los stakeholders. Un puente para el desarrollo práctico de la ética empresarial y de la responsabilidad social corporativa. *VERITAS* 2 (17): 205–224.
———. 2012. El modelo de empresa plural desde el enfoque de los skateholders. *Debats* 46: 71–80.
———. 2016. El desarrollo de la ética empresarial ante los avances de la neurociencia organizacional y la neuroética. *Pensamiento: Revista de investigación e Información Filosófica* 72 (273): 921–940. https://doi.org/10.14422/pen.v72.i273.y2016.009.
Gordon, Nancy. 2013. Women and Leadership: An Integrative Focus on equality. *Journal of Interdisciplinary Feminist Thought* 7 (1): Article 11.
Habermas, Jürgen. 1998. *The Inclusion of the Other: Studies in Political Theory.* Cambridge: Massachusetts Institute of Technology.
Herreros, Carlos. 2012. *#Neuro-Management.* Madrid: LID Editorial Empresarial.
Hickman, Gill R. 1998. *Leading organizations: Perspectives for a new era.* Thousand Oaks: Sage.

Hon, Linda Childers, and Brigitta Brunner. 2000. Diversity Issues and Public Relations. *Journal of Public Relations Research* 12 (4): 309–340. https://doi.org/10.1207/S1532754XJPRR1204_2.
Illouz, Eva. 2007. *Intimidades congeladas*. Buenos Aires: Katz Editores.
———. 2010. *La salvación del alma moderna*. Madrid: Katz Editores.
Kabat-Zinn, John. 1994. *Wherever You Go, There You Are: Mindfulness Meditation in Everyday Life*. New York: Hyperion Books.
———. 2005. *Guided Mindfulness Meditation*. Sounds True.
Kossek, Ellen Ernst, and Sharon A. Lobel. 1996. *Managing Diversity: Human Resource. Strategies for Transforming the Workplace*. Cambridge: Blackwell.
Kounios, John, and Mark Beeman. 2009. The Aha! Moment: The Cognitive Neuroscience of Insight. *Current Directions in Psychological Science* 18 (4): 210–216. https://doi.org/10.1146/annurev-psych-010213-115154.
Laval, Christian, and Pierre Dardot. 2013. *La nueva razón del mundo. Ensayo sobre la sociedad neoliberal*. Barcelona: Gedisa.
Lieberman, Matthew D. 2009. The Brain's Braking System (and How to 'Use Your Words' to Tap Into It). *Neuroleadership* 2: 9–14.
Medina-Vicent, Maria. 2017a. Beyond National Borders: Business Ethics and Gender Equality in a Globalized Economy. Comparing the EU, Spain, and Brazil. In *Economic Behaviour. Economy, Business and People*, ed. Alezandra Zbuchea, Constantin Brătianu, and Florina Pînzaru, 169–200. Newcaslte upon Tyne: Cambridge Scholars Publishing.
———. 2017b. Responsabilidad Social de Género y obligatoriedad moral. *Femeris. Revista Multidisciplinar de Estudios de Género* 2 (1): 30–48.
Medina-Vicent, Maria, and Sonia Reverter-Bañón. 2016. La perspectiva de género como una mirada crítica al mundo. *Asparkía. Revista de Investigació Feminista* (29): 11–16. https://doi.org/10.6035/Asparkia.2016.29.1.
Muñoz-Rodríguez, David, and Antonio Santos-Ortega. 2017. Las cárceles del capital humano: trabajo y vidas precarias en la juventud universitaria. *Recerca. Revista de Pensament i Anàlisi* 20: 59–78. https://doi.org/10.6035/Recerca.2017.20.4.
Ngunjiri, Faith Wambura. 2010. Lessons in Spiritual Leadership from Kenyan Women. *Journal of Educational Administration* 48 (6): 755–768. https://doi.org/10.1108/09578231011079601.
Norton, J. Renae, and Ronald E. Fox. 1997. *The Change Equation: Capitalizing on Diversity for Effective Organization Change*. Washington, DC: American Psychological Association.
Ochsner, Kevin N., and James J. Gross. 2005. The Cognitive Control of Emotion. *Trends in Cognitive Sciences* 9 (5): 242–249. https://doi.org/10.1016/j.tics.2005.03.010.
Ochsner, Kevin N., Silvia A. Bunge, James J. Gross, and John D.E. Gabrieli. 2002. Rethinking Feelings: An FMRI Study of the Cognitive Regulation of Emotion. *Journal of Cognitive Neuroscience* 14 (8): 1215–1229.
Ochsner, Kevin N., Rebecca D. Ray, Jeffrey C. Cooper, Elaine R. Robertson, Sita Chopra, John D.E. Gabrieli, and James J. Gross. 2004. For Better or for Worse: Neural Systems Supporting the Cognitive Down- and Up- Regulation of Negative Emotion. *NeuroImage* 23 (2): 483–499. https://doi.org/10.1016/j.neuroimage.2004.06.030.
Pallarés-Domínguez, Daniel. 2015. Hacia una conceptualización dialógica de la neuroeducación. *Participación Educativa* 4 (7): 133–141. https://doi.org/10.4438/1886-5o97-PE.
Papalini, Vanina. 2006. La subjetividad disciplinada: de la contracultura a la autoayuda. In *La comunicación como riesgo: cuerpo y subjetividad*, ed. V. Papalini, 21–79. La Plata: Ediciones Al Margen.
Ringleb, Al H., and David Rock. 2009. The emerging field of neuroleadership. *Neuroleaderhsip Journal*. https://neuroleadership.com/portfolio-items/the-emerging-field-of-neuroleadership-2/ Accessed 13 Sept 2018.
Rock, David. 2009. *Your Brain at Work. Strategies for Overcoming Distraction, Regaining Focus, and Working Smarter All Day Long*. New York: HarperBusiness.
———. 2010. Leadership on the Brain. *Harvard Business Review*. https://hbr.org/2010/04/leadership-on-the-brain. Accessed 13 Sept 2018.

Rock, David, and Heidi Grant. 2016. Why Diverse Teams Are Smarter. *Harvard Business Review*. https://hbr.org/2016/11/why-diverse-teams-are-smarter. Accessed 13 Sept 2018.

Rock, David, and Linda J. Page. 2009. *Coaching with the Brain in Mind: Foundations for Practice*. New York: Wiley.

Rock, David, and Jeffrey Schwartz. 2006. The Neuroscience of Leadership. *Strategy+Business*. https://www.strategy-business.com/article/06207. Accessed 13 Sept 2018.

———. 2010. Impacting Leadership with Neuroscience. *Strategy & Business*. https://www.strategy-business.com/article/06207?gko=6da0a. Accessed 13 Sept 2018.

Rock, David, Heidi Grant, and Jacqui Grey. 2016. Diverse Teams Feel Less Comfortable and That's Why They Perform Better. *Harvard Business Review*. https://hbr.org/2016/09/diverse-teams-feel-less-comfortable-and-thats-why-they-perform-better. Accessed 13 Sept 2018.

Rosenthal, Seth A., and Todd L. Pittinsky. 2006. Narcissistic Leadership. *The Leadership Quarterly* 17 (6): 617–633. https://doi.org/10.1016/j.leaqua.2006.10.005.

Rost, Joseph C. 1991. *Leadership for the Twenty-First Century*. Santa Barbara: Praeger.

———. 1993. Leadership Development in the New Millennium. *Journal of Leadership Studies* 1 (1): 91–110. https://doi.org/10.1177/107179199300100109.

———. 1995. Leadership: A Discussion About Ethics. *Business Ethics Quarterly* 5 (1): 129–142. https://doi.org/10.2307/3857276.

———. 1997. Moving from Individual to Relationship: A Postindustrial Paradigm of Leadership. *Journal of Leadership Studies* 4 (4): 3–16. https://doi.org/10.1177/107179199700400402.

———. 1998. Leadership and Management. In *Leading Organizations: Perspectives for a New Era*, ed. Gill Robinson Hickman, 93–124. Thousand Oaks: Sage.

Rost, Joseph C., and Richard A. Barker. 2000. Leadership Education in Colleges: Toward a 21st Century Paradigm. *Journal of Leadership Studies* 7 (1): 3–12. https://doi.org/10.1177/107179190000700102.

Smith, Khali, and David Rock. 2018. Approaching Diversity with the Brain in Mind. *Strategy+Business*. https://www.strategy-business.com/article/Approaching-Diversity-with-the-Brain-in-Mind. Accessed 13 Sept 2018.

Smith, Wendy K., Marianne W. Lewis, and Michael L. Tushman 2016. "Both/and" Leadership. *Harvard Business Review*. https://hbr.org/2016/05/both-and-leadership. Accessed 13 Sept 2018.

Susaeta, Lourdes, and José Emilio Navas. 2006. La diversidad cultural como fuente de ventaja competitiva. Una aplicación a la empresa multinacional. *Cuadernos de Estudios Empresariales* 15: 153–177.

Thomas, R. Roosevelt, Jr. 1990. From Affirmative Action to Affirming Diversity. *Harvard Business Review* 68 (2): 107–117.

Triandis, Harry C. 2003. The Future of Workforce Diversity in International Organisations: A Commentary. *Applied Psychology* 52 (3): 486–495. https://doi.org/10.1111/1464-0597.00146.

Turnbull, Martin. 2015. A Model of Motivation for Facilitating Sustainable Change. *Neuroleadership Journal*. https://membership.neuroleadership.com/material/a-model-of-motivation-for-facilitating-sustainable-change-vol-5/. Accessed 13 Sept 2018.

Western, Simon. 2012. *Coaching and Mentoring. A Critical Text*. London: Sage.

———. 2013. *Leadership: A Critical Text*. London: Sage.

Wexler, B. 2011. Neuroplasticity, Culture, and Society. In *The Oxford Handbook of Neuroethics*, ed. Judy Illes and Barbara J. Sahakian, 743–760. Oxford: Oxford University Press.

Chapter 13
Moral Neurolearning by Machines: Artificial Values, Intelligences and Neural Networks

Patrici Calvo

13.1 Introduction

Artificial Intelligence (AI) is booming as never before. The initial development of this technology took place in the summer of 1956, when ten researchers interested in neural networks held a six-week workshop to try "(…) to discover how to make machines that use language, form abstractions and concepts, solve problems so far reserved to human beings, and improve themselves" (Bostrom 2014). However, during the 1980s, AI's target audience all but succumbed to a lack of interest, as it gradually dawned on both its potential developers and consumers that the advances predicted and the expectations generated for AI's practical applications were, in most cases, overly optimistic (Bostrom 2014). However, in the last decade there has been a major resurgence of AI thanks, above all, to the emergence of neuroscience, and studies in this field have provided a better understanding of neural learning processes. Furthermore, the consolidation and expansion of the Fourth Industrial Revolution, which has had an important impact on sectors as disparate as retail, cosmetics, communication, education and health care, has revealed how the potential of these sectors can be achieved by optimizing the use of scarce resources and improving production, communication, education, health care and decision-making processes.

Neuroscience is a branch of physiology that studies the cerebral dimension of human or animal behaviour by using invasive and non-invasive techniques to capture images of brain activity, such as functional Magnetic Resonance Imaging (fMRI), Repetitive Transcranial Magnetic Stimulation (rTMS), magnetoencephalography (MEG) and Positron Emission Tomography (PET). Since its emergence at the end of the 1990s, the main aim of this field of research has been to explore and analyse the cerebral processes underpinning human behaviour, especially those

P. Calvo (✉)
Universitat Jaume I, Castellón de la Plana, Spain
e-mail: calvop@uji.es

P. Calvo, J. Gracia-Calandín (eds.), *Moral Neuroeducation for a Democratic and Pluralistic Society*, https://doi.org/10.1007/978-3-030-22562-9_13

involved in decision-making processes across the diverse contexts of human activity. The improved knowledge of the brain and the huge quantity of data produced by neuroscientific studies have also led to considerable progress in AI. Neuroscience has, firstly, made it possible to improve the design of artificial neural networks and, secondly, it has provided a seminal source of Big Data on the emotions, feelings, values and other factors and elements involved in the learning and decision-making processes employed by intelligent machines.

With regard to the Fourth Industrial Revolution, this refers to the digital transformation process that was begun by the European car industry during the first decade of the twenty-first century, and which was subsequently applied to other areas of human activity (Calvo 2017, 2018, 2019a, b, c; Zamagni 2018). Two factors have underpinned this process of digital transformation, Key Enabling Technologies (KETs) and the convergence of the following three technological areas: the Internet of Things (IoT), Big Data (BD) and Artificial Intelligence (AI). The combination of these provides the potential for the *hyperconnectivity*, *datafication* and *algovernance* of everything: machines, people, animals and processes. The IoT makes it possible to recreate a cyber-physical space capable of digitally *hyperconnecting* things, processes, people and animals. Big Data converts, compiles and stores the huge amounts of data created by *digital hyperconnectivity*. Finally, AI processes and transforms Big Data into relevant information and then into knowledge that can be applied in making rational decisions, which in turn make it possible to optimize and improve all the processes involved in a specific area and organization. The incredible results provided by the digital transformation of sectors such as industry, retail and health care, and the specific application of AI as a part of this, has given rise to an unprecedented interest in AI's potential for improved development and application among academics, researchers, professionals and investors.

One of the most relevant concerns currently being raised in the field of AI regards algorithmic governance in politics and business. The expectations currently being generated about the benefits of the application of algorithmic governance are so high—the eradication of corruption, nepotism and selfishness in decision-making processes; optimization of resources; greater competitiveness; increased profits, etc.—that management and/or elected representatives are being replaced by mathematical decision-making models. For example, some companies have already included algorithms with the right to speak and vote on their boards of directors, and there are countries where people have been allowed to choose an algorithm as their political representative. Outstanding cases include Michihito Matsuda, the robot created with a female appearance, which came third in the elections for mayor of Tama New Town (Tokyo) in April 2018; and SAM (Semantic Analysis Machine), the algorithm being prepared for the New Zealand's 2020 general election (Calvo 2019a, c).

Nonetheless, processing Big Data about people's preferences, opinions and habits with regard to a specific activity in order to generate statistics is one thing; actually governing such areas is a wholly different matter. Among other things, this is because governance requires an understanding of autonomy, prudence, justice, the common good, recognition, tolerance and responsibility, and currently decision-making algorithms are incapable of displaying even the slightest competence or capacity for moral learning and development. Indeed, Artificial Moral Intelligence (AMI) faces many problems, above all concerned with three main areas: firstly, moral competence, for example, the possibility of developing an artificial structure capable of making judgements in response to value conflicts, experiential learning or moral decision-making; secondly, moral capacity, such as how, when and why machines should use an artificial moral structure to resolve value conflicts or make fair decisions; and, thirdly, moral learning, which concerns the possibility of a machine creating, learning, assimilating and developing critical, post-conventional moral knowledge.

Together these three issues are an obstacle AI has yet to overcome. Moral conflicts are problematic and demand moral plurality, discursive processes, values with a universal scope, post-conventional points of view, constant criticism, individualized case analysis and the choice of possible courses of action, whereas the current model of moral machine learning, or Moral Artificial Intelligence (MAI) is dilemmatic, descriptive, complete, inconsistent, empirical, decisionist, positivist and vicarious. As Diego Gracia has stated, dealing with moral controversy in a dilemmatic way leads to unreal situations and immoral decisions (Gracia 2000). For this reason, work is ongoing, although with little success to date, on the moral competences and capacities of the application of AI.

The task of this study is twofold, to undertake a critical review of the current potential for moral competence, capacity and learning by AI applications, and to offer guidelines for their improved future development. Firstly, consideration is given to how neuroscience has contributed to AI's current success, which in turn provides a foundation to consider the different models of moral learning that have been proposed for AI. Finally, an analysis is undertaken of the elements of AI that at present prevent it from successfully becoming an AMI.

13.2 Artificial Neural Networks (ANN): The Possibility That *Things* Can Learn

At the beginning of the twenty-first century, with the technological convergence of the internet, the web 2.0 and digital devices, and especially the mass use of social media, like Facebook, and mobile phones, Anabel Quan-Haase and Barry Wellman described Computer Mediated Communication (CMC) in terms of hypoconnectivity and as "the availability of people for communication anywhere and anytime" (2005: 285). As José Luis Orihuela has argued, this trend has generated a new communication ecosystem that enables, among other significant factors, "(…) the understanding and visualization of social relationships and modes of information distribution" (Llavina 19 November 2015). Most importantly, it has initiated a continuous flow of data, created by the multiple interactions of this ecosystem's users, and this has in turn given rise to the potential analysis, management and strategic use of this data across diverse areas, such as economics, advertising and politics. As José Van Dijck argues:

> Connectivity did not take long to develop into a valuable resource as engineers did not take long to find ways of coding all this information into algorithms to contribute to shaping a particular form of online society (…). The biggest, most influential platforms – Facebook, Twitter, You Tube and LinkedIn – underwent a real explosion in their number of users and their potential to generate money (…) The result of the interconnection of these platforms was the emergence of a new infrastructure: an ecosystem of connected media (…). This move from network communication to a society shaped by platforms and from a participatory culture to a true culture of connectivity happened in short space of no more than ten years (van Dijck 2016: 29).

Today, thanks to the development of the IoT, this trend has reached still higher levels of sophistication through what has come to be known as *digital hyperconnectivity*—the propensity and possibility of people to remain permanently connected with *everything* they consider important, i.e. not just people—through a constant flow and exchange of data over the internet.[1] *Digital hyperconnectivity* involves the recreation of cyber-physical spaces[2] which, structured around virtually interconnected software, devices and sensors, allow the digital connectivity of *everything*—things, processes, animals and people (IoT).[3] It also involves the implementation of algorithms that enable the governance of the processes involved, as well as both the *things* and people that are connected (AI). Finally, it offers the possibility of extending the mathematical models with the necessary objective, relevant information, whereby they can make rational decisions more effective, optimize the processes involved, and improve the behavioural predictability of the *things connected* (Serpanos and Wolf 2017).

Neuroscience plays a key role in these developments, especially with regard to the improved performance of functions that underpin the IoT's application to a range of areas of human activity. Among other things, a variety of neuroscientific studies have become a key part of AI development (Nguyen 2017). They provide important data and information about the way the brain functions within different contexts of human activity and during decision-making processes, which has led to

---

[1] What keeps us connected is data, not language. In other words, in the era of digitalisation, communication is understood as data exchange, which makes it possible to stay connected with any user and/or element of the cyber-physical ecosystem.

[2] Cyber-physical spaces consist of at least five basic characteristics: connection, generation, accumulation, prediction and decision.

[3] As Effy Vayena and Urs Gasser (2016: 27) explain: "Our increasing interaction with digital technologies and devices, and their effects on us and our behavior, have given birth to new concept: the digital phenotype." Since first being proposed in 2015 (Jain et al. 2015), this concept is closely linked to the idea of an *extended phenotype* suggested by Richard Dawkins (1982), who argued that the idea of the phenotype should not be limited to biological processes. Thereby, interactions with the environment and the way we alter it also come into play, generating a much broader phenotype. See, Vayena and Gasser, "Capturing and understanding these interactions allows greater understanding of how we function" (2016: 27).

improvements in the internal design of mathematical models used in AI, and also bolstered and optimized AI's predictive and decision-making capacities.

The greater knowledge provided by the latest neuroscientific advances on human behaviour has generated an unprecedented development of AI. Still more recently, machine learning algorithms, based on sets of rules that enable autonomous decisions to be made as responses to practical problems, have now developed into *deep learning* (Bostrom 2014; Iba 2017), which consists of algorithms based on artificial neural networks (ANN). These are computational models, such as *Google's Neural Machine Translation* (NMT), that imitate the operation of the nervous system in living organisms. However, these ANNs do not make decisions based on a closed model in which possible strategies and responses have been previously established.[4] Instead, the algorithm itself learns through experience, for example, through continuous trial and error processes, and it establishes or strengthens connections between parameters or between the different layers of processing units (artificial neurons that specialize in analysing a specific characteristic of the object observed) in order to improve its abstractions and perceptions of reality and, in this way, achieve the intended results (Conde et al. 2018; Hirasawa et al. 2018; Webb 2018; Higham and Higham 2018).

Furthermore, neuroscience has also become an important and inexhaustible source of data, which serves to improve the predictability of connected things and rationalize the decision-making processes of algorithmic models. For example, studies of the different neuroscientific subdisciplines, such as neuroeconomics, neuroethics, neuromarketing and neuropolitics, are continually offering large quantities of objective, measurable data on human behaviour. As data directly sourced from human brains, this can be processed by Big Data analysis tools to minimize any uncertainty underlying the cyber-physical system through the improvement of predictions and decisions made using mathematical models (Chen and Lin 2014; Higham and Higham 2018; Marblestone et al. 2016).

Undoubtedly, the alliance between the neurosciences and the technological convergence consisting of Big Data, the IoT and AI is changing the modus operandi of a range of sectors (Celik 2016). As Oriuela argues, the huge predictive potential underlying the computational analysis of major databases, "(...) is not only a treasure trove for marketing, it also has wider consequences that include the early detection of illnesses, the mapping of epidemics, planning urban transport logistics, managing crime prevention, predicting market behaviour and, of course, the study of public opinion beyond user surveys" (Llavina 19 November 2015). However, the consequences of these changes are not always positive, or acceptable to most of those affected by the system (O'Neil 2016; Donati 2019). Political and economic neuromarketing, for example, can use cyber-physical ecosystems to design and implement highly addictive campaigns capable of modulating or manipulating subjects' free will within a system, as well as diminishing or inhibiting voters, leaders

[4] This approach is also known as supervised learning, as the machine requires a degree of human support or help in order to make decisions, such as, for example, in knowing what is right and what is wrong.

and customers' capacity for critical judgement, in order to maximize the individual profit of a few.[5] In addition, political and economic institutions are able to monitor the behaviour—including the private behaviour—of voters, customers and competitors in order to predict and control their actions and decisions. Furthermore, they can create attractive and/or addictive products and prompt decision-making processes based on this data.

In regard to these developments, what is of particular importance is the gathering and use of Big Data related to the emotions and its use to understand and shape subjects' behaviour in a practical context. The mass leak of at least 87 million Facebook users' private data by the consultancy firm Cambridge Analytica and its fraudulent use as a tool for influencing both the US presidential elections and the United Kingdom's 2016 referendum on European Union membership (Brexit) (Sumpter 2018) are good examples of this. Another is the United States' National Security Agency's (NSA) mass espionage programme uncovered in 2013 by the whistle-blower Edward Snowden. Using a range of Big Data analysis tools, digital hyperconnectivity and algorithms based on AI, the NSA committed 2776 violations of privacy laws during the first 12 months of its implementation (Saiz 16 August 2013). Furthermore, there is the replacement of people in democratic governments and companies by applications based on mathematical models that make rational decisions based on probability and which, although generally beneficial to the organization from an economic point of view, are unjust and irresponsible with regard to the many real people who may work at a firm, or who are affected by the actions or decisions (O'Neil 2016) taken. It may therefore be argued that a key problem faced by AI is the development of machines' moral competence, capability and learning. To achieve this, over the last few years different moral learning models have been developed for 'things', which are based on the latest neuroscientific advances.

13.3 Moral Learning by Machines: The Possibility That *Things* Could Acquire Values

One of the issues that most concerns contemporary AI developers is the possibility of machines making autonomous, rational, moral decisions. The implementation of driverless cars, for example, has been hindered because of the difficulty in defining a moral reference for these vehicles when making decisions in the event of a conflict with the values underlying the practice of intelligent mobility (Monasterio 2017). On the one hand, this concerns decisions where human lives are at stake, such as whether it is prudent to change course to avoid running someone over, even if the manoeuvre could put the passengers' lives at risk. The question remains to be

[5] For an in-depth study of the ethical aspects of neuromarketing, see Conill (2012) and Conill and García-Marzá (2012).

resolved whether developers should choose the greatest good for the greatest number of people and implement a balanced compromise between an excessive action and the failure to act, or else establish a reciprocal recognition of human dignity as a moral decision-making criterion. On the other hand, the question remains as to whether it is possible to introduce an axiological framework, whereby rational AI decision-making could be guided, while at the same time learning from experience.

As Nick Bostrom (2014) has indicated, at present there are five main approaches being explored in order to address the problem of how to introduce values that would enable the creation of an AI structure capable of issuing judgements about value conflicts, learning from experience and taking moral decisions.

(a) *Learning through evolutionary selection*: this involves search algorithms that emulate the biologically evolved process of the acquisition of values. The algorithms, firstly, expand the catalogue of values and, secondly, carry out tasks to select the best value and eliminate the worst one—in other words, the ones that function best and worst in the real world. The use of this dual process is comparable to the biologist John Maynard Smith's (1982) definition of an evolutionary stable strategy (ESS) and replication dynamic (RD), which is concerned solely with values and not social behaviour strategies. Therefore, the search algorithm would, on the one hand, compile the values from the real world used to resolve moral conflicts that remain unchanged over time (expansion), while, on the other, using an evaluation function (contraction), it would establish an order of materiality to eliminate those values scoring worst in a test. Finally, it would process the available information in order to either mute or specify both the final values and optimal ones in order to achieve its objectives. The problem with this kind of assimilation of values is that, as Bostrom argues, "Nature can be a great experimenter, but it would never pass an ethics exam—it contravenes the Helsinki declaration and all rules of moral decency in all senses" (Bostrom 2014: 188). Furthermore, it is a conventional, strategic proposal, based on instrumentalizing values. Any one criterion for the choice and acquisition of values is not linked to better guidance for taking action in any one specific case, but rather to a better adaptation to the medium involved. The lack of any critical dimension throughout this process also generates a purely conventional axiological framework, which can serve to encourage and perpetuate injustices.
(b) *Learning through reinforcement:* this involves an algorithm receiving some kind of recompense, evaluation, stimulus or information (from the programmer or the environment, for example) about the decisions it has made. To do this, it is assigned an evaluation function or utility which, through experience (trial and error), allows the algorithm to refine its estimates of the practical use of certain values and then associate them with responses and behaviours, in order to maximize the benefit obtained, as well as reinforce or correct its decisions. For example, giving a reward (a point, for example) for each decision made on the basis of certain values, but without telling it why, or what the values are.

However, as Samuel Bowles states, the main problem with this kind of learning by reinforcement lies in the fact that good incentives do not make good citizens (2016). As has been shown in field studies and laboratory experiments with strategy games, over time the beneficiaries of incentives become incapable of acting morally if there are no reasonable expectations that they will be rewarded for it. In addition, behaving morally does not always lead to intrinsic or extrinsic positive recompense for people. Quite often, doing what should be done has a high cost, such as poorer adaptation, less personal welfare or reduced profits. The problem is that human beings not only have a moral structure, they also have a capacity to criticise the conventional, and discern what is fair through dialogue and consensus with all those affected.

(c) *Learning through association:* this is a matter of associating values with concepts, things, processes, experiences and objectives, as happens when, for example, human beings confer a set of values on concepts such as a person, profession, institution, etc. In the case of AI, it would be a matter of selecting, linking and adapting complex values to specific objectives; in other words, governing with leadership, respect, honesty and responsibility. The problem in this case is raised by the difficulty of achieving an in-depth knowledge of the mechanism that permits human beings to associate values with concepts, which is linked to a neurocognitive architecture that can only be applied through a complete brain emulation (current AI is a very limited emulation of the human brain). Firstly, there is a lack of clarity about the reference to be used by AI, either for determining the values that should be associated with a particular concept, or for criticising them or, if thought to be necessary, changing them for others. Secondly, there is a further lack of clarity regarding the reasoning used by AI for choosing, associating and allowing itself to be guided by these values without the reinforcement provided by a utility function. All this seems to indicate that, as in the previous approaches, the reference used once again consists of the values that enable the given objective to be successfully achieved and a concern to maximize the benefits provided by this achievement. As a result, learning by association once again gives rise to the instrumentalization of values.

(d) *Learning through scaffolding:* this would involve establishing methods for selecting an AI's objectives, purposes, motivations and rules. Firstly, it would involve providing an AI with an instructional scaffolding consisting of provisional instrumental objectives, as well as final objectives that would serve as a frame of reference for achieving its instrumental objectives. Yet this instructional scaffolding must be sufficiently simple to grant the AI the possibility of developing it or, if it sees fit, modifying it. Secondly, it would involve providing a motivational scaffolding to ensure that the AI developed provisional objectives and goals. However, while learning through scaffolding is an especially interesting line of enquiry, its implementation is currently curtailed by the lack of any solid foundation for its development. At present, the limitations include the potentially excessive control and influence wielded by the programmer, which is contrary to modern notions of AI. In this case, although the primary frame of reference is simple and open so that it can be developed and modified by the AI, it is enriched through the experience provided by the AI's pursuit of its provisional aims and motivations, which as a rule are introduced by the programmer. Therefore, the AI has to be equipped with a learning structure containing a minimum of content that can be modified on the basis of its experience and in response to its growing capacity to deal with issues of greater complexity.

(e) *Learning through evaluation:* this involves providing an AI with criteria that enable it to autonomously learn the values it is intended to use, and also define the best way to act in accordance with them on the basis of estimates. Unlike learning through scaffolding, this approach does not change the AI's aim, but rather the evaluation it makes in order to achieve its aim. On the one hand, the problem raised by this approach lies both with the criterion to be used, and the possibility of developing a sufficiently evolved AI; one that can detect and comprehend the axiological structure of its environment by elaborating increasingly precise hypotheses, which in turn depend on estimates based on its own principles and the empirical data available. While on the other hand, as with the other proposals, the issue is how "(…) to ensure that the AI feels motivated to pursue the values described in the way we want it to" (Bostrom 2014: 197). In other words, the fact that the AI knows and interprets the axiological framework perfectly does not mean it will follow it, and in addition, its estimates of values and objectives are based on what the AI is, not what it should be.

Ultimately, as Bostrom states, "The engineering of objectives is not an established discipline. At the moment, we do not know how to transfer human values to a digital computer, even one with a human level of AI" (Bostrom 2014: 207). Yet, the current expectations regarding AI would require an unprecedented development of its moral dimension in order to achieve them. Among other important issues, it is necessary to find the right AI design and to provide it with effective specifications in order to morally educate it, whereby a transition can be made to AMI, and thereby modern challenges, such as algorithmic governance or autonomous transport, can be addressed. However, given the current state of technology this is by no means a realistic objective, above all with regard to AI's limited capacity to engage with the idiosyncrasy and complexity of a moral environment.

13.4 The Moral Education of Machines: Concerning the Non-existence of Morally Intelligent Technology

Over the last few decades, different neuroscientific studies have provided quantitative and qualitative information about, among other things: the motivational heterogeneity of the economic agent; the different neurobiological basis for individual and cooperative behaviour; the important role played by *prosocial emotions* and the moral judgements involved in rational decision-making processes; the capacity of emotions emanating from moral values to modulate self-interested behaviour; the links between behaviour and areas of the brain related to cognition and moral judgement; and the capacity of formal and informal education to affect everything.[6]

With regard to the possible role of education in the emergence and promotion of moral thinking, moral neuroeducation, as a neuroscientific discipline concerned with the moral learning process and its application in teaching regarding different areas of activity,[7] such as education and business,[8] has proposed guidelines for the theoretical and practical development of learning processes based on the brain, as well as techniques and strategies for their normative and practical implementation.[9]

Firstly, mirror neurons are responsible for a significant part of human behaviour, even in the first months of life (Warneken and Tomasello 2006). For the purpose of adaptation, they emotionally connect the brains of those who belong to a specific society in order to generate behavioural mimesis through the evaluation of direct experience. Within, this vicarious learning process, feelings and emotions play a fundamental role as they allow members of a society to better adapt themselves through unconscious learning processes, which are much quicker than conscious ones that use imitation and empathy (Rizzolatti et al. 2001). As Giácomo Rizzolatti, Leonardo Fogassi and Vittorio Gallese argue in "Neurophysiological mechanisms underlying the understanding and imitation of action" (2001):

> Mirror neurons are a particular class of visuo-motor neurons, originally discovered in a sector of the ventral premotor cortex of monkeys, called area F5. Area F5 is characterized by the presence of neurons that code goal-related motor acts, such as hand and mouth grasping. Some of these cells are motor neurons, others also respond to visual stimuli. Some of them are activated by the presentation of three-dimensional objects, whereas others—mirror neurons—require action observation for their activation. The main functional characteristic of mirror neurons is that they become active both when the monkey makes a particular action (for example, when grasping an object or holding it), and when it observes another individual (monkey or human) making a similar action. Typically, mirror neurons do not respond to the sight of a hand mimicking an action in the absence of the target. Similarly, they do not respond to the observation of an object alone, even when it is of interest to the monkey.

Secondly, the human brain is tremendously malleable. In other words, the nervous system has the capacity to change the structure and function of the brain in order to continually adapt to its environment, experience and stimuli by strengthening or weakening the synapses connecting neurons, and through neurogenesis. For education, this neural plasticity implies the possibility of being able to influence the cognitive development of any student to improve their capacities and eliminate

[6] For a study of all these contributions, see Calvo (2018).

[7] For a detailed study of moral neuroeducation, see Pallarés-Domíguez (2016, 2017)

[8] For a detailed study of moral neuroeducation in business, see Medina-Vicent and Pallarés-Domínguez (2017).

[9] For a detailed study of these and other contributions, see González-Esteban (2016).

learning disorders that hider the development of their abilities or make learning impossible (García 2008).

Thirdly, emotions have a cognitive basis and are extremely important in all learning, reasoning and decision-making processes (Damasio 2003). As Antonio Damasio argues, positive emotions encourage memorization and learning, while negative emotions make the transfer of information between the different areas of the brain and its proper processing more difficult.

Fourthly, the brain is capable of establishing bridges between the knowledge acquired and new information by repeating what it wishes to assimilate: values, information, guidelines, behaviours, etc. A key result of this is that it allows for learning to be optimised and it fosters critical reflection among students. The brain's capacity for bridge-building is, among other things, due to the automation of mental processes, which reduces the consumption of working memory, and, as has been shown, individuals with a greater available working memory are better at learning and critical reflection (Willingham 2011: 144–150).

Finally, formal education makes it possible to affect normative and practical aspects by raising awareness, and strengthening and promoting the values, guidelines and behaviours we have good reason to appreciate. It also reviews or eradicates those that have partially or entirely lost their meaning and legitimacy. Based on these five contributions from neuroeducation, one can, at least, shed light on the *non-place* of AMI. Among the most notable issues raised by the different approaches to the learning and acquisition of values there are five that must be underscored as of special importance: moral structure, content, paternalism, motivation and intelligence.

Moral structure: the current proposals for machine moral learning are above all concerned with the application, development and implementation of a part of AI's moral structure—the mechanisms for the acquisition of values—, but little attention has been paid to the other important and necessary parts of the human moral structure. Nothing is said, for example, about how to establish a mechanism that emulates moral judgements regarding what is fair and unfair, nor the capacity to criticise both knowledge and action. And likewise, no consideration is given to the possibility of generating prosocial emotions that lead to values being followed regardless of the benefits derived from their implementation. In this sense, many of the current approaches establish a kind of substitute for artificial moral judgements, which is based on a utility function related to their specific aims, as opposed to an adherence to principles, values or rules. The consequences of this, derived from their application, could be accepted by all those affected through a practical discourse based on certain logical rules and a procedural moral principle (Cortina 1993: 208).

Moral content: current approaches to moral machine learning are barely concerned with moral content, or how to specify, justify and enrich the moral values, which enable the machine's structure to operate in a proper, legitimate manner. An impression is given that this disinterest lies in a lack of confidence in current AMI, which in turn provides the basis for the excessive paternalism observed in most approaches. It is striking that AlphaGo's autonomous, creative capacity to design and apply strategies never before thought of by human beings has been the object of praise, yet AMI's creative capacity is controlled by patches to prevent the existing values being enriched with arguments never before put forward by human beings. It seems that there is a clear acknowledgement of both the impossibility of approaching moral complexity on the basis of mathematical formalism, as well as the mismatch between computer rationality and practical rationality. Finally, a high level of mistrust in the possibility of leaving the future direction of humanity in the hands of AMI is acknowledged.

Moral paternalism: in order to speak of AMI, firstly, AI has to have a capacity for freedom as well as autonomy in order to grant it the capacity of making laws for itself. However, underlying this is a form of paternalism, a paternalistic freedom or autonomy. The latter is largely due to the lack of confidence in AMI's capacity to both sustain its pre-established objectives (cerebral puncture) and to establish the necessary line of reasoning that would allow itself to be guided by a framework of moral values. The general lack of confidence in AI compels programmers to insert patches to prevent a possible cerebral puncture, as well as utility functions that are based on aggregating scores for intermediate objectives and maximizing benefits to prevent any possible abandonment of the AI's objectives. Thereby, this promotes an almost scrupulous adherence to objectives and rules.

Moral motivation: in human beings, the motivation to want to acquire and follow moral principles, values and norms stems from the duty to comply with what we have good reason to consider to be fair and desirable. However, AI builds a motivational structure based on to what extent different principles, values and rules are involved in achieving its given objectives through a pre-established utility function, or via programmers' estimates concerning values. Thereby, the decision-making criteria of most techniques for acquiring and learning values is the maximization of profit, and as a result AMI instrumentalizes values to improve results. Furthermore, within AMI there is an important underlying deficit of emotional reasoning, which hinders its establishment and development.

Moral intelligence: one of the biggest problems faced by the development of AMI, particularly with regard to moral learning, is the concept of managing intelligence. AI's intelligence is understood as the capacity to achieve complex objectives by analysing empirically quantifiable regulations. Although it is narrow, reductionist and false, this point of view is the principal mantra of moral machine learning studies. The reduction of moral machine learning to the possibility of compiling, analysing, rating and ordering values impedes its development and, more seriously, can give rise to and generate highly immoral decisions and consequences. A complex objective might be conceived, such as in the film *Minority Report*, as predicting, finding and arresting or killing possible criminals. From a moral point of view not considered by AMI, the word intelligent derives from *intelligentia*, which is taken from the term *eligantina*—elegance—which, in turn, comes from the term *eligĕre*—to choose. Based on this etymology, in the Classical World, the word *intelligentia* was linked to character and ethics, and was used to define people who were capable of overcoming circumstances, barring the occasional impulse, to plan a fully human life for themselves in relation to others through the taking of fair, prudent decisions (Ortega y Gasset 1967: 13–14).

Therefore, while human intelligence may be related to ethics, in so far as practical knowledge aspires to universality in order to be able to clarify, justify and rationally guide actions and decisions across all spheres of activity, the current approach to ethics in AI development is linked to the merely empirical analysis of facts in terms of the observation, evaluation and organization of circumstances, as well as, opinions regarding the sensory world. The selection of the moral values that form AI's axiological framework is based on quantifiable empirical relations, which are based on an empirical analysis of a society's behavioural facts and opinions. However, this analysis does not consider these values intrinsic value and capacity to guide actions and decisions in a fair, congratulatory sense. Therefore, AMI not only reveals a significant underlying deficit of emotional reason, but also of communicative reason, which hinders or renders still more impossible AI's foundation and development.

Finally, the notion of AI underpinning the different approaches to value acquisition and learning is conventional and strategic, and it is also based on an acritical observation of the patterns, opinions and behaviours of a specific, or global or nomological community based on programmers' estimates of what is fair and good. Yet, the problem arises with regard to the fact that moral values cannot be conventional, nor can they be instrumentalized. Moral values are post-conventional and, therefore, as Adela Cortina states, their roots are anchored in intersubjectivity (2010). The fact remains that, as Margaret A. Boden, one of the greatest experts on AI, has stated, "Thinking that you can resolve conflicts as human as the Middle East (and, while you are at it, Northern Ireland or Catalonia) with AI is utterly ridiculous. Those who believe in *singularity* do not know the limits of modern AI. They are basing themselves only on exponential technological advances, while they are ignoring one fact: the increased power of computers and the availability of data does not guarantee human-standard AI" (Corazón 16 April 2018).

References

Bostrom, Nick. 2014. *Superintelligence: Paths, Dangers, Strategies*. Oxford: Oxford University Press.

Bowles, Samuel. 2016. *The Moral Economy: Why Good Incentives Are No Substitute for Good Citizens*. New Haven/London: Yale University Press.

Calvo, Patrici. 2017. Ética de las cosas ante el desafío de la Industria 4.0. *Corresponsables*. http://www.corresponsables.com/actualidad/etica-en-la-industria. Accessed 14 Sept 2018.

———. 2018. *The Cordial Economy – Ethics, Recognition and Reciprocity*. Cham: Springer.

———. 2019a. Democracia algorítmica: consideraciones éticas sobre la datafícación de la esfera pública. *Reforma y Democracia* 74: 5–30.

———. 2019b. Etificación, la transformación digital de lo moral. *Kriterion. Revista de filosofia* 60: 144.

———. 2019c. Neuroaprendizaje moral de las máquinas en la política y la economía. In *Neuroeducación moral y Democracia*, ed. Jesús Conill and Domingo García-Marzá. Granada: Comares.

Chen, Xue-Wen, and Xiaotong Lin. 2014. Big Data Deep Learning: Challenges and Perspectives. *IEEE Access* 2: 514–525. https://doi.org/10.1109/ACCESS.2014.2325029.

Conde, Aintzane, Ander Arriandiaga, José Antonio Sanchez, Eva Portillo, Soraya Plaza, and Itziar Cabanes. 2018. High-Accuracy Wire Electrical Discharge Machining Using Artificial Neural Networks and Optimization Techniques. *Robotics and Computer-Integrated Manufacturing* 49: 24–38.

Conill, Jesús. 2012. Neuroeconomía y Neuromarketing. ¿Más allá de la racionalidad maximizadora? In *Guía Comares de Neurofilosofía Práctica*, ed. Adela Cortina, 39–64. Granada: Comares.

Conill, Jesús, and Domingo García-Marzá. 2012. Neuroeconomía y Neuromarketing. In *La comunicación empresarial y la gestión de los intangibles en España y Latinoamérica*, ed. Justo Villafañe, 143–163. Madrid: Pearson.

Cortina, Adela. 1993. *Ética aplicada y democracia radical*. Madrid: Tecnos.

———. 2010. *Justicia cordial*. Madrid: Trotta.

Corazón, Alvaro. 2018, 16 April. Margaret Boden: El mayor obstáculo para el desarrollo de la IA es que no sabemos qué hace un cerebro humano. *Yorokubu*. https:// www.yorokobu.es/margaret-a-boden/. Accessed 14 Sept 2018.

Damasio, Antonio R. 2003. *Looking for Spinoza: Joy, Sorrow and the Feeling Brain*. New York: Harvest Book Harcourt, INC.

Dawkins, Richard. 1982. *The Extended Phenotype: The Gene as the Unit of Selection*. Oxford: Oxford University Press.

Donati, Pierpaolo. 2019. *Sociología relacional de lo humano*. Barañain: EUNSA.

García, Emilio. 2008. Neuropsicología y educación. De las neuronas espejo a la teoría de la mente. *Revista de Psicología y Educación* 1 (3): 69–90.

González-Esteban, Elsa. 2016. El desarrollo de la ética empresarial ante los avances de la neurociencia organizacional y la neuroética. *Pensamiento* 72 (273): 921–940.

Gracia, Diego. 2000. La deliberación moral. El papel de las metodologías en ética clínica. In *Comités de Ética Asistencial*, ed. José Sarabia and Manuel de los Reyes, 21–41. Madrid: Asociación de Bioética Fundamental y Clínica.

Higham, Catherine F., and Desmond J. Higham 2018. Deep Learning: An Introduction for Applied Mathematicians. eprint arXiv:1801.05894.

Hirasawa, Toshiaki, Kazuharu Aoyama, Tetsuya Tanimoto, Soichiro Ishihara, Satoki Shichijo, Tsuyoshi Ozawa, Tatsuya Ohnishi, Mitsuhiro Fujishiro, Keigo Matsuo, Junko Fujisaki, and Tomohiro Tada. 2018. Application of Artificial Intelligence Using a Convolutional Neural Network for Detecting Gastric Cancer in Endoscopic Images. *Gastric Cancer* 21: 653–660. https://doi.org/10.1007/s10120-018-0793-2.

Iba, Hitoshi. 2017. *Evolutionary Approach to Machine Learning and Deep Neural Networks: Neuro*. Cham: Springer.

Jain, Sachin H., Brian W. Powers, Jared B. Hawkins, and John S. Brownstein. 2015. The digital phenotype. *Nature Biotechnology* 33: 462–463. https://doi.org/10.1038/nbt.3223; https://www.ncbi.nlm.nih.gov/pubmed/25965751

Llavina, Xantal. 19 de noviembre de 2015. Ahora los medios compiten con sus audiencias. *El Periódico*. https://www.elperiodico.com/es/entrevistas-talento-digital/20151119/entrevista-jose-luis-orihuela-4686719. Accessed 14 Sept 2018.

Marblestone, Adam H., Greg Wayne, and Konrad P. Kording. 2016. Toward an Integration of Deep Learning and Neuroscience. *Frontiers in Computational Neuroscience* 10: 94), 1–94),41.

Maynart Smith, John. 1982. *Evolution and the Theory of Game*. Cambridge: Cambridge University Press.

Medina-Vicent, Maria, and Daniel Pallarés-Domínguez. 2017. Las huellas del Neurosexismo en la literatura popular del *management* dirigida a mujeres. *Política y Sociedad* 54 (3): 683–705. https://doi.org/10.5209/POSO.53074.

Monasterio-Astobiza, Aníbal. 2017. Ética algorítmica: Implicaciones éticas de una sociedad cada vez más gobernada por algoritmos. *Dilemata. Revista internacional de éticas aplicadas* 24: 185–217.

Nguyen, Anh M. 2017. *AI Neuroscience: Visualizing and Understanding Deep Neural Networks*. Laramie: University of Wyoming.

O'Neil, Cathy. 2016. *Weapons of Math Destruction: How Big Data Increases Inequality and Threatens Democracy*. New York: Broadway Books.

Ortega y Gasset, José. 1967. *The Origin of Philosophy*. New York: Northon & Company.

Pallarés-Dominguez, Daniel. 2016. *Bases neuroéticas para la educación moral: una neurorracionalidad dialógica y práctica*. Castellón: Publicacions de la Universitat Jaume I.

———. 2017. Neuroeducación en diálogo: neuromitos en el proceso de enseñanza aprendizaje y la educación moral. *Pensamiento* 72 (273): 941–958.

Quan-Haase, Anabel, and Barry Wellman. 2005. *How Computer-Mediated Hyperconnectivity and Local Virtuality Foster Social Networks of Information and Coordination in a Community of Practice*. In International Sunbelt Social Network Conference. Redondo Beach, CA.

Rizzolatti, Giácomo, Leonardo Fogassi, and Vittorio Gallese. 2001. Neurophysiological Mechanisms Underlying the Understanding and Imitation of Action. *Nature Reviews Neuroscience* 2 (9): 661–670.

Saiz, Eva. 13 de agosto de 2013. La NSA infringió las normas de privacidad en miles de ocasiones. *El País*. https://elpais.com/internacional/2013/08/16/actualidad/1376631278_378738.html. Accessed 14 Sept 2018.

Serpanos, Dimitrios, and Marilyn Wolf. 2017. *Internet-of-Things (IoT) Systems: Architectures, Algorithms, Methodologies*. Cham: Springer.

Sumpter, David. 2018. *Outnumbered: From Facebook and Google to Fake News and Filter-Bubbles – The Algorithms That Control Our Lives*. New York: Bloomsbury.

van Dijk, Jose. 2016. *La cultura de la conectividad: una historia crítica de las redes sociales*. Buenos Aires: Siglo XXI Editores.

Vayena, Effy, and Urs Gasser. 2016. Strictly Bioomedical? Stretching the Ethics of the Big Data Ecosystem in Biomedicine. In *The Ethics of Biomedical Big Data*, ed. Brent D. Mittelstadt and Luciano Floridi, 17–40. Cham: Springer.

Warneken, Felix, and Michael Tomasello. 2006. Altruistic Helping in Human Infants and Young Chimpanzees. *Science* 311 (5765): 1301–1303.

Webb, Sara. 2018. Deep Learning for Biology. *Nature* 554: 555–557. https://doi.org/10.1038/d41586-018-02174-z.

Willingham, Daniel. 2011. *¿Por qué a los niños no les gusta ir a la escuela? Una respuesta de un neurocientífico al funcionamiento de la mente y sus consecuencias en el aula*. Barcelona: Graó.

Zamagni, Stefano. 2018. Civil Economy. A New Approach to the Market in the Age of the Fourth Industrial Revolution. *Recerca. Revista de Pensament i Anàlisi* 23:151–168.

Index

P. Calvo, J. Gracia-Calandín (eds.), *Moral Neuroeducation for a Democratic and Pluralistic Society*, https://doi.org/10.1007/978-3-030-22562-9

The manufacturer's authorised representative in the EU is Springer Nature Customer Service Centre GmbH, Europaplatz 3, 69115 Heidelberg, Germany. If you have any concerns regarding our products, please contact ProductSafety@springernature.com

Printed and bound by CPI Group (UK) Ltd, Croydon, CR0 4YY
05/10/2026
02228131-0005